Advancement of Selective Laser Melting by Laser Beam Shaping

Vom Promotionsausschuss der
Technischen Universität Hamburg

zur Erlangung des akademischen Grades
Doktor-Ingenieur (Dr.-Ing.)

genehmigte Dissertation

von
Tim Marten Wischeropp

aus
Neumünster

2021

1. Gutachter: Prof. Dr.-Ing. Claus Emmelmann
2. Gutachter: Prof. Milan Brandt

Tag der mündlichen Prüfung: 21. September 2021

Light Engineering für die Praxis

Reihe herausgegeben von

Claus Emmelmann, Hamburg, Deutschland

Technologie- und Wissenstransfer für die photonische Industrie ist der Inhalt dieser Buchreihe. Der Herausgeber leitet das Institut für Laser- und Anlagensystemtechnik an der Technischen Universität Hamburg sowie die Fraunhofer-Einrichtung für Additive Produktionstechnologien IAPT. Die Inhalte eröffnen den Lesern in der Forschung und in Unternehmen die Möglichkeit, innovative Produkte und Prozesse zu erkennen und so ihre Wettbewerbsfähigkeit nachhaltig zu stärken. Die Kenntnisse dienen der Weiterbildung von Ingenieuren und Multiplikatoren für die Produktentwicklung sowie die Produktions- und Lasertechnik, sie beinhalten die Entwicklung lasergestützter Produktionstechnologien und der Qualitätssicherung von Laserprozessen und Anlagen sowie Anleitungen für Beratungs- und Ausbildungsdienstleistungen für die Industrie.

Weitere Bände in der Reihe https://link.springer.com/bookseries/13397

Tim Marten Wischeropp

Advancement of Selective Laser Melting by Laser Beam Shaping

Springer Vieweg

Tim Marten Wischeropp
Institute of Laser and System Technologies
(iLAS)
Hamburg University of Technology
Hamburg, Germany

ISSN 2522-8447 ISSN 2522-8455 (electronic)
Light Engineering für die Praxis
ISBN 978-3-662-64584-0 ISBN 978-3-662-64585-7 (eBook)
https://doi.org/10.1007/978-3-662-64585-7

Responsible Editor: Alexander Gruen
This Springer Vieweg imprint is published by the registered company Springer-Verlag GmbH, DE part of Springer Nature.
The registered company address is: Heidelberger Platz 3, 14197 Berlin, Germany

Abstract

Selective Laser Melting (SLM) offers significant advantages for the manufacturing of complex, high-quality parts. Especially in the aerospace, automotive, medical and tooling industries, SLM is becoming more and more popular. However, its market share is still small compared to conventional manufacturing technologies. Major drawbacks hindering an industrial ramp-up are low productivity, high part costs and issues with quality and reproducibility. Comprehensive research has been done to overcome these challenges, but little attention has been paid to addressing them by optimizing the laser beam profile. Therefore, this thesis aims to examine the potential of an optimized laser beam profile to increase the productivity and stability of the SLM process.

For this, a numerical model is developed to investigate the effect of the beam profile on the melt pool dynamics that are difficult to examine experimentally. Furthermore, experiments are performed with a Gaussian and a donut beam profile to study the effect of the beam profile on the manufacturing of single tracks and the multi-layer SLM process for AlSi10Mg and AlSi16Sc. The investigations show that there is a significant effect of the beam profile on the SLM process and that a donut beam profile offers a series of advantages over the "standard" Gaussian beam profile. Due to the more homogeneous energy input of the donut beam profile, the vaporization in the process is reduced by an average of 40% and the magnitude of the thermo-capillary flow is lowered by an average of 20% for the process parameters investigated. This leads to a stabler melting process with fewer fluctuations in the process and a reduction in spatters of 35%. Furthermore, the occurrence of keyhole porosity is significantly reduced at high energy inputs. This increases the feasible energy input and process window size, which makes the process more robust against variations and potentially increases the long-term stability. Due to the mentioned effects, the build rate can be increased by 60% and the costs reduced by up to 15%. In addition, experiments with the high-strength material AlSi16Sc suggest that the effect of the beam profile is even more pronounced for materials that are difficult to process due to the tendency to form gas or keyhole porosity in SLM.

The results demonstrate significant benefits offered to the SLM process by an optimized beam profile. However, further research is necessary to establish beam shaping for SLM in the industry. Foremost, additional investigations have to be done to examine the physical limits of the process experimentally and to explore the effect of beam shaping for other materials. Moreover, alternative beam shaping techniques that meet industrial requirements by increasing the robustness of the beam profile should be studied.

Table of Contents

Nomenclature

Symbol	Description	Unit
α	Accommodation coefficient	
β_R	Recondensation rate	
γ_{al}	Interfacial tension at air-liquid interface	N/m
γ_{as}	Interfacial tension at air-solid interface	N/m
γ_{ls}	Interfacial tension at liquid-solid interface	N/m
γ_s	Surface tension	N/m
δ	Thermal diffusivity	m^2/s
ε	Emissivity/Absorptivity of the surface	
ε_t	Turbulent energy dissipation	kg^2m^2/s^2
θ	Dynamic wetting angle	rad; °
θ_0	Equilibrium wetting angle	rad; °
θ_{0_Al}	Equilibrium wetting angle AlSi10Mg	rad; °
λ	Characteristic dimension for porous media	m
μ	Dynamic viscosity of fluid	kg/(m·s)
μ_{Al}	Dynamic viscosity of AlSi10Mg	kg/(m·s)
μ_S	Dynamic viscosity of fluid during solidification	kg/(m·s)
μ_{TL}	Dynamic viscosity of fluid at liquidus temperature	kg/(m·s)
ν	Kinematic viscosity	m^2/s
ν_t	Turbulent kinematic viscosity	m^2/s
ρ	Density	kg/m^3
σ	Surface tension	N/m^2
σ_{Al}	Surface tension of AlSi10Mg	N/m^2
σ_B	Stefan Boltzmann constant	W/m^2K
ϕ	Surface-irradiation angle of laser beam	rad; °
χ	Vaporization energy per atom	J/mol
∇	Nabla Operator	1/m

Symbol	Description	Unit
ΔH_{F_Al}	Latent heat of fusion of AlSi10Mg	J/kg
ΔH_V	Latent heat of vaporization	J/kg
ΔH_{V_Al}	Latent heat of vaporization of AlSi10Mg	J/kg
Δx_{min}	Minimum cell size	mm
a	Packing density of powder	%
a1	Packing density of power determined by com. methods	%
a_{Al_l}	Absorptivity of liquid AlSi10Mg	%
a_{Al_s}	Absorptivity of solid AlSi10Mg	%
a_{fres}	Fresnel absorption coefficient	%
a_{fres_i}	Local fresnel absorption coefficient	%
a_{occ}	Occupancy rate of the machine	%
a_{sim}	Packing density of powder in numerical model	%
b	Build rate	mm^3/s
b_{GST}	Build rate of standard process parameters for AlSi10Mg	mm^3/s
b_{MAX_D}	Maximum feasible *b* with donut beam (AlSi10Mg)	mm^3/s
b_{MAX_D2}	Maximum feasible *b* with donut beam (AlSi16Sc)	mm^3/s
b_{MAX_G}	Maximum feasible *b* with Gaussian beam (AlSi10Mg)	mm^3/s
b_{MAX_G2}	Maximum feasible *b* with Gaussian beam (AlSi16Sc)	mm^3/s
br	Actual build rate (considering jump vectors and delays)	mm^3/s
$br_{G,D}$	Build rate for Gaussian or donut beam profile	mm^3/s
C_{build}	Costs of build job	€
C_h	Hourly machine rate	€
C_{inv}	Investment costs for machine	€
C_m	Maintenance costs	€
C_{Mat}	Costs of material for build job	€
C_{Mat_kg}	Costs of material per kg	€
C_{oh}	Overhead costs	%
C_p	Specific heat capacity	J/kgK

C_{p_Al}	Specific heat capacity of AlSi10Mg	J/kg·K
C_{Pro}	Costs of process for build job	€
C_v	Vapor specific heat at constant volume	J/kg·K
C_{v_Al}	Specific heat capacity of vapor AlSi10Mg	J/kg·K
d_f	Diameter of laser beam in focus	µm
d_l	Laser beam diameter	µm
E_A	Total amount of absorbed energy	J
E_{Al}	Young's modulus of AlSi10Mg	N/m^2
E_R	Total amount of reflected energy	J
$\vec{f}$	Momentum source vector	kg/m^2s^2
$\overrightarrow{f_c}$	Capillary force vector	kg/m^2s^2
$\overrightarrow{f_d}$	Specific drag force	kg/m^2s^2
$\overrightarrow{f_p}$	Vapor pressure force vector	kg/m^2s^2
$\overrightarrow{f_{tc}}$	Thermo-capillary force vector	kg/m^2s^2
F	Fluid fraction coefficient	
F_{cap}	Capillary force	N/m
F_D	Drag force coefficient	kg/m^3s
F_S	Fluid fraction source term	1/s
F_{VAP}	Vapor pressure force	N
$\vec{g}$	Gravity vector	m/s^2
h	Specific enthalpy	J/kg
h_b	Specific heat flux of single laser beam ray	W/m^2
h_{build}	Height of build job	mm
h_{ce}	Convective heat transfer coefficient	W/m^2K
h_{ce_Al}	Convective heat transfer coefficient of AlSi10Mg	W/m^2K
h_E	Specific energy source term	W/m^3
h_l	Levelling height of baser plate / layer thickness	µm
h_{lb}	Energy input by laser beam	W/m^2

h_{lb_D}	Specific energy of donut laser beam profile	W/m^2
h_{lb_G}	Specific energy of Gaussian laser beam profile	W/m^2
h_{l_pow}	Layer thickness on top of a previous molten layer	μm
h_{l_sim}	Layer thickness used in numerical model	μm
h_s	Hatch distance	μm
h_v	Vaporization energy loss	W/m^2
K	Permeability	$1/m^2$
K_1	Vaporization pressure constant	
K_j	Factor for jump vectors and laser-scanner delays	
k	Thermal conductivity	W/(m·K)
k_{Al}	Thermal conductivity of AlSi10Mg	W/(m·K)
k_f	Thermal conductivity of the fluid	W/(m·K)
k_t	Turbulent kinetic energy	$kg \cdot m^2/s^2$
k_μ	Temperature-dependent viscosity gradient	kg/K/m/s
l_c	Characteristic length for heat conduction	mm
l_{c_mp}	Characteristic length (width of melt pool)	μm
l_{mp}	length of cylindrical segment / melt pool	mm
LED	Linear Energy Density	J/mm
M	Molecular weight of vapor	kg/mol
m_S	Mass source term	kg/m^3s
$\vec{n}$	Normal vector	
p_0	Specific ambient gas pressure	Pa/m
Pe	Peclet number	
P_L	Laser power	W
$P_{L\_MAX\_D}$	Maximum feasible P_L for donut beam (AlSi10Mg)	W
$P_{L\_MAX\_D2}$	Maximum feasible P_L for donut beam (AlSi16Sc)	W
$P_{L\_MAX\_G}$	Maximum feasible P_L for Gaussian beam (AlSi10Mg)	W
$P_{L\_MAX\_G2}$	Maximum feasible P_L for Gaussian beam (AlSi16Sc)	W
p_r	Specific vapor pressure	Pa/m

p_{sat}	Specific saturation pressure	Pa/m
q_{con}	Convective energy loss	W/m^2
q_{rad}	Radiative energy loss	W/m^2
q_{vap}	Vaporization energy loss	W/m^2
r	Distance from the beam center	µm
r_b	Radius of laser beam	µm
S	Solid fraction	
s	Surface curvature	1/m
S_C	Critical solid fraction	
S_{CO}	Coherent solid fraction	
SED	Surface Energy Density	J/mm^2
t_{build}	Total time of build job	s
t_{dep}	Depreciation time of machine	year
t_{melt_t}	Total melting time of build job	s
t_{rec}	Recoating time per layer	s
t_{rec_t}	Total recoating time of build job	s
T	Temperature	K
T_b	Boiling temperature of fluid	K
T_{b_Al}	Boiling temperature of AlSi10Mg	K
T_{BP}	Preheating temperature of base plate	K
T_C	Temperature at critical solid fraction	K
T_{CO}	Temperature at coherent solid fraction	K
T_{L_Al}	Liquidus temperature of AlSi10Mg	K
T_{LT}	Liquidus temperature	K
T_{LT_Al}	Liquidus temperature of AlSi10Mg	K
T_{MAX}	Maximum surface temperature	K
T_S	Surface temperature	K
T_{S_Al}	Solidus temperature of AlSi10Mg	K
T_{ST}	Solidus temperature	K

T_{ST_al}	Solidus temperature of AlSi10Mg	K
T_V	Temperature of void (surrounding gas)	K
T_{V1}	Saturation temperature of void (surrounding gas)	K
V_{build}	Total volume of parts in build job	mm^3
v_m	Velocity of melt	mm/s
v_{ms}	Mean surface melt velocity	m/s
v_S	Scanning speed	mm/s
$v_{S\_MAX\_D}$	Maximum feasible v_S for donut beam (AlSi10Mg)	mm/s
$v_{S\_MAX\_D2}$	Maximum feasible v_S for donut beam (AlSi16Sc)	mm/s
$v_{S\_MAX\_G}$	Maximum feasible v_S for Gaussian beam (AlSi10Mg)	mm/s
$v_{S\_MAX\_G2}$	Maximum feasible v_S for Gaussian beam (AlSi16Sc)	mm/s
VED	Volumetric Energy Density	J/mm^3
VED_{MIN}	Threshold VED to avoid LoF porosity (AlSi10Mg)	J/mm^3
VED_{MIN_2}	Threshold VED to avoid LoF porosity (AlSi16Sc)	J/mm^3
VED_{MOD}	Modified Volumetric Energy Density	J/mm^3
VED_{MOD_MAX}	Threshold VED_{MOD} to avoid keyhole porosity	J/mm^3
$VED_{MOD\_MAX\_D}$	Threshold VED_{MOD_MAX} for donut beam (AlSi10Mg)	J/mm^3
$VED_{MOD\_MAX\_D2}$	Threshold VED_{MOD_MAX} for donut beam (AlSi16Sc)	J/mm^3
$VED_{MOD\_MAX\_G}$	Threshold VED_{MOD_MAX} for Gaussian beam (AlSi10Mg)	J/mm^3
$VED_{MOD\_MAX\_G2}$	Threshold VED_{MOD_MAX} for Gaussian beam (AlSi16Sc)	J/mm^3
w	Width	µm
w_{mp}	Width of cylindrical segment / melt pool	mm
x_{min}	Minimum cell size	mm
y_{1-5}	Positions of laser beam evaluated in the numerical model	mm
y_e	End position of laser beam in numerical model	mm
y_s	Starting position of laser beam in numerical model	mm
z	Levelling height of building platform	µm
z_f	Focus position relative to working plane	mm

Abbreviations

2D	Two dimensional
3D	Three dimensional
μCT	Micro Computer Tomography
ALE3D	Arbitrary Lagrangian-Eulerian Technique
AM	Additive Manufacturing
AVG	Average
CAD	Computer Aided Design
CFD	Computational Fluid Dynamics
CLIP	Continuous Liquid Interface Production
CT	Computer Tomography
DEM	Discrete Element Method
DMLS	Direct Metal Laser Sintering
EBM	Electron Beam Melting
EDX	Energy-dispersive X-ray spectroscopy
FAVOR	Fractional Area-Volume Obstacle Representation
FCRC	Flight Crew Rest Compartment
FDM	Fused Deposition Modeling
FEM	Finite Element Method
FVM	Finite Volume Method
IAPT	Institution for Additive Manufacturing Technologies
LB	Lattice Boltzmann
LBM	Laser Beam Melting
LMD	Laser Metal Deposition
LMF	Laser Metal Fusion
LoF	Lack-of-fusion
LPBF	Laser Powder Bed Fusion
Ma	Marangoni number
MP	Melt pool

n/a	Not applicable
PDE	Partial Differential Equation
Pe	Peclet number
PF	Phase field
PLIC	Piecewise Linear Interface Construction
R&D	Research and development
Re	Reynolds number
RNG	Renormalization-Group Method
ROI	Region of interest
SEM	Scanning Electron Microscope
SLM	Selective Laser Melting
STD	Standard deviation
TEM	Transverse Electromagnetic Mode
US	United States of America
VOF	Volume of Fluid
w/o	Without

1 Introduction

This chapter gives an overview of the background, motivation and approach of this thesis.

1.1 Background

Manufacturing companies are facing a transforming business environment that causes several challenges for them. On the one hand, customers show an increasing demand for individualized products and short lead times. On the other hand, rising competition, especially from fast-growing markets such as China, India and parts of Latin America, are putting pressure on established companies [107, 158, 159]. The fact that costs for commodities and energy are rising reinforces the latter [222]. Furthermore, the rising environmental awareness in the population increases the demand for resource-efficient production technologies [202, 241]. One can expect that the ecological footprint of a manufactured product will play a considerable role in the choice of suppliers in the future [93, 241]. Therefore, highly flexible, digitalized and automatized manufacturing capabilities, as well as resource-efficient production processes, can be seen as critical success factors for manufacturing companies in the future [66, 107, 200, 202].

While traditional production processes are limited in flexibility and cost competitiveness for individualized products, based on the economy of scale, Additive Manufacturing (AM) shows a different behavior. It requires no special tooling and can be highly automated, based on its digital nature. Furthermore, AM is resource-efficient, as the overall energy consumption and amount of material waste (usually $< 5\%$) is lower compared to traditional manufacturing technologies [161]. The fact that AM addresses the challenges of today's manufacturing companies can be seen as one of the fundamental reasons for the remarkable growth of the AM market in recent years, as shown in Figure 1-1 [240]. Despite the growth, the AM market is still small compared to the overall machine tool market, where the annual revenue was ~14 trillion US$, compared to ~10 billion US$ of the AM market in 2018 [112, 240].

T. M. Wischeropp, *Advancement of Selective Laser Melting by Laser Beam Shaping*,
Light Engineering für die Praxis, https://doi.org/10.1007/978-3-662-64585-7_1

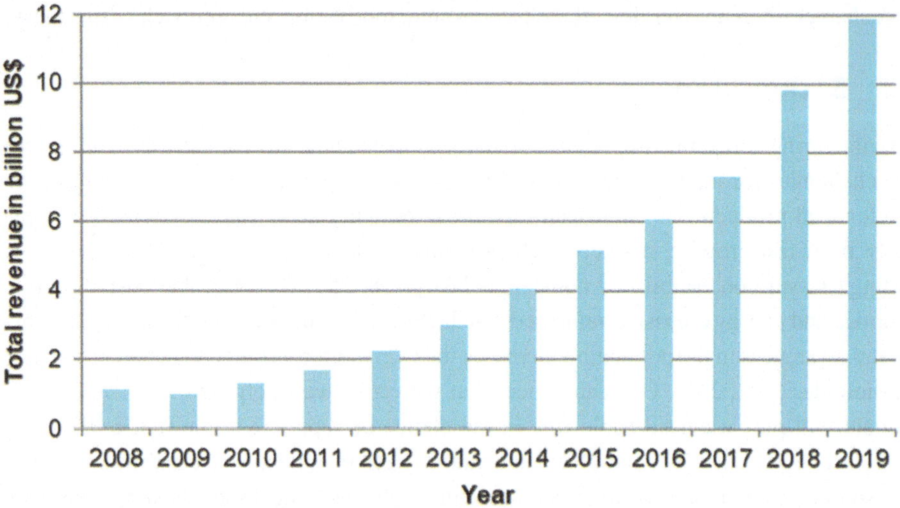

Figure 1-1: Total annual revenue of the AM market from 2008 to 2019 (based on [240]).

For metal AM, *Selective Laser Melting (SLM)* is the technology with the highest rate of industrialization [10, 240]. The technology shows an increasing number of industrial applications due to its capability to manufacture highly complex parts with very good mechanical properties [10, 240]. The most relevant markets for the SLM technology are the aerospace (22%), automotive (6%), medical (15%) and machine tool markets (20%)[1] [10]. Furthermore, most companies invest in the technology intending to manufacture final parts [240]. Figure 1-2 shows some examples of successful industrial applications of SLM.

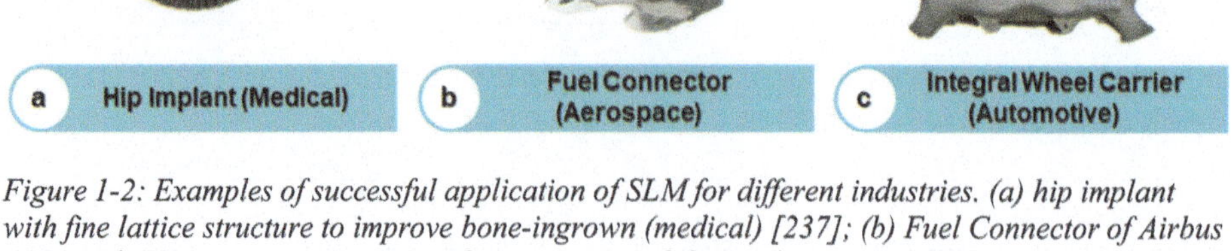

Figure 1-2: Examples of successful application of SLM for different industries. (a) hip implant with fine lattice structure to improve bone-ingrown (medical) [237]; (b) Fuel Connector of Airbus A380 with 50% cost saving compared to conventional design (aerospace) [70]; (c) Integral wheel carrier with > 35% weight saving compared to conventional design (automotive, developed by Fiat Chrysler and Fraunhofer IAPT) [151].

Significant investments of several companies underpin the commitment of the industry to the technology. As an example, General Electric invested over $599 million (USD) to

[1] based on systems sales in 2018

acquire controlling shares of the German machine manufacturer Concept Laser GmbH in 2016 and opened an AM Centre in Lichtenfels, Germany for $150 million (USD) in 2019 [63]. Furthermore, the German machine manufacturer SLM Solutions Group AG announced their largest single order of machines by an unnamed Asian company, with an order volume of 60 machines ($56.5 million (USD)) in 2017 [211] and the SLM service provider BeamIT announced the purchase of 15 SLM machines at the beginning of 2020 [213].

Ongoing trends to further increase the maturity of the SLM technology include the development of machines with higher productivity and larger built chambers as well as the increase in digitalization and automation [22, 212, 240, 252]. As an example, the size of the build chambers increased by a factor of 7 from 2012 to 2020 (SLM250 HL compared to Concept Laser X2000) and larger machines have already been announced for the near future [64, 152, 196]. Additionally, machines that combine SLM with post-processing (e.g., milling, heat treatment, etc.) are becoming more and more popular [188, 196, 240].

1.2 Motivation

To further increase the application of the SLM technology in the industry, it is crucial to leverage the technology's potential for serial production. Currently, this is limited, among other things, due to low process speeds that lead to high part costs and issues with reproducibility [10, 22, 243]. However, the demand for high process speeds and reproducibility are contradictory requirements. On the one hand, high process speeds require a high energy input that increases vaporization in the process. On the other hand, an increase in vaporization enhances the instabilities in the process and provokes defects such as porosity [69, 137]. In addition, the laser welding mode changes to the undesired keyhole mode at a specific threshold energy input [35, 205]. Hence, the maximum process speed for one laser is limited by the threshold energy input that still allows a stable melting process.

Consequently, productivity improvements are mainly achieved by increasing the number of lasers, and advancements in reproducibility are primarily driven by incremental improvements of the process (e.g., optimized gas stream in the build chamber, improved powder characteristics, etc.) [10, 240]. Furthermore, research to increase the productivity of each single laser beam is focused on the optimization of the process parameters, such as the laser power, scanning speed, hatch distance or layer thickness [3, 74, 128, 247]. Currently, little attention is paid to the influence of the laser characteristics (e.g., laser beam profile, wavelength, polarization, …), although it is known that they are essential in laser material processing [82, 101]. Furthermore, the limited research done on this topic primarily investigates the effect of the beam profile on single tracks or microstructure and not on the productivity or stability of the SLM process [186, 190, 265].

In today's SLM machines, lasers with a Gaussian beam profile are used that are characterized by a high intensity in the center of the beam [190, 210]. It can be assumed that beam profiles with a more even energy distribution, such as a donut beam profile (see Figure 4-3 on page 56), could reduce the vaporization for a given set of process parameters and hence increase the threshold energy at which the process becomes unstable [62, 186]. Based on this prerequisite, higher process speeds and potentially a stabler SLM process would be feasible with alternative laser beam profiles. This assumption is supported by investigations done on other laser material processes, such as laser welding or laser surface hardening, where alternative beam profiles have already demonstrated advantages over a Gaussian beam profile [84, 219, 230]. Therefore, the underlying research question of this thesis is: Can an alternative laser beam profile, such as a donut beam profile, enhance the productivity and process stability of the SLM process?

1.3 Approach

Several investigations (see Figure 1-3) are carried out to answer the research question.

Based on the state of the art, a numerical micro-scale model is developed to investigate the effect of the beam profile on melt pool characteristics that are difficult to examine experimentally, such as the fluid flow and magnitude of vaporization. With the help of the numerical model, the melting of single tracks is simulated for two different beam profiles: a standard Gaussian beam profile and a donut beam profile that is characterized by a more homogeneous energy input. Afterward, single-track and density cube experiments are carried out on a modified SLM machine with a Gaussian and a donut beam profile. The single-track experiments are performed to get a basic understanding of the effect of the laser beam profile on the appearance, size, and shape of single tracks. The density cube experiments are carried out to investigate the effect of the beam profile on the feasible energy input[2], process window, build rate, and spatter formation of the multi-layer SLM process. The results of the experiments are discussed and analyzed with numerical results. Furthermore, the industrial relevance of the outcomes is discussed and recommendations for further investigations are provided.

The investigations are primarily performed for the aluminum alloy AlSi10Mg as it is the most commonly used aluminum alloy in SLM. However, additional multi-layer SLM experiments are performed with a high strength aluminum AlSi16Sc to examine if the effect of the beam profile is influenced by the alloy composition.

It must be noted that the research focuses on the effect of the beam profile (Gaussian and donut beam profiles) on the characteristics of the melting process which limit the productivity and stability of SLM such as vaporization, the occurrence of defects, and spatters. Investigations on the effect of the beam profile on other characteristics such as microstructure, surface roughness, residual stresses or mechanical properties are not part of the thesis and should be done in future work.

---

[2] In the context of this thesis, the **feasible energy input** is defined as the maximum energy input that still avoids the generation of keyhole porosity (< 0.5% keyhole porosity).

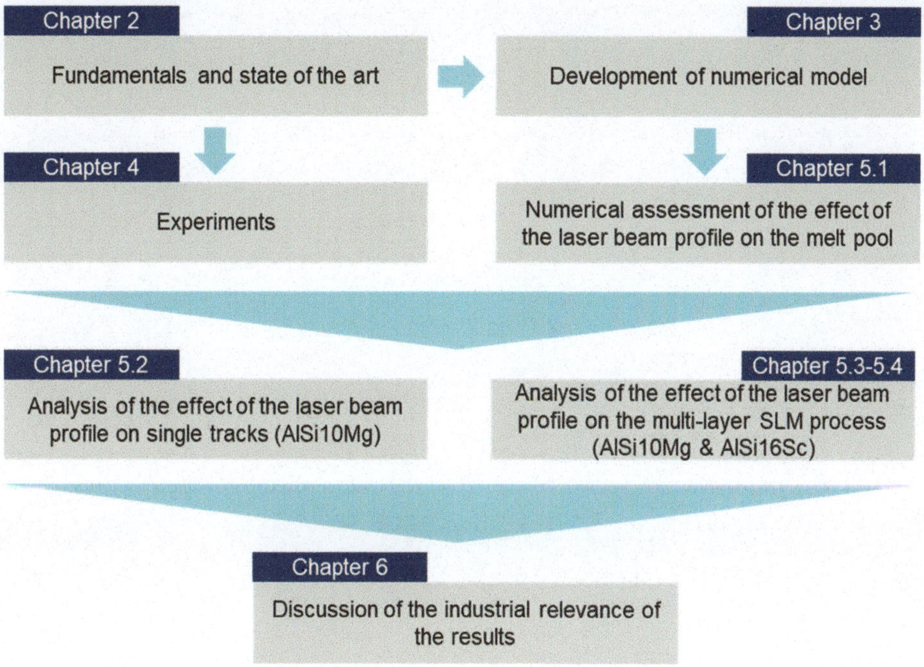

Figure 1-3: Structural approach of the thesis.

2 Fundamentals

The following chapter provides an overview of the fundamentals and the state of the art that are relevant for the understanding of this thesis. In the beginning, an overview of Additive Manufacturing (AM) and, in particular, the Selective Laser Melting (SLM) is given (Chapter 2.1 and 2.2). Afterwards, the state of the art of simulation of the SLM process is presented (Chapter 2.3) and an outline of the application of laser beam shaping in SLM and related processes is given (Chapter 2.4).

2.1 Additive Manufacturing

In Germany, manufacturing processes are divided into six categories as defined by the DIN 8580 and Additive Manufacturing is considered a subgroup of the first category *primary forming* [58]. In the Anglo-American region, on the other hand, manufacturing processes are divided into three categories, with AM being one of the main categories, as shown in Figure 2-1 [65].

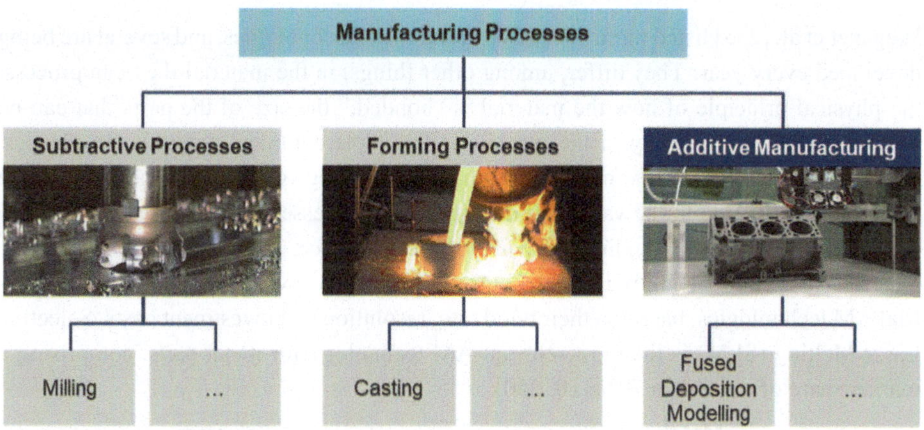

Figure 2-1: Classification of manufacturing processes in the Anglo-American region [65].

The fundamental principle of AM is the creation of a 3D object by successive manufacturing and merging of layers (see Figure 2-2). The layer-wise manufacturing principle breaks down a complex 3D manufacturing process into several, much simpler, 2D manufacturing steps [65]. This is the foundation of many advantages that AM has over conventional manufacturing technologies. The 2D manufacturing steps are much easier to automate and do not require special tooling. This makes AM very flexible and highly suitable for individualized products and mass-customization [65, 202]. In addition to this, the layer-wise manufacturing principle enables the production of highly complex structures with internal features and undercuts that are very expensive or infeasible with other manufacturing technologies [22, 228].

T. M. Wischeropp, *Advancement of Selective Laser Melting by Laser Beam Shaping*,
Light Engineering für die Praxis, https://doi.org/10.1007/978-3-662-64585-7_2

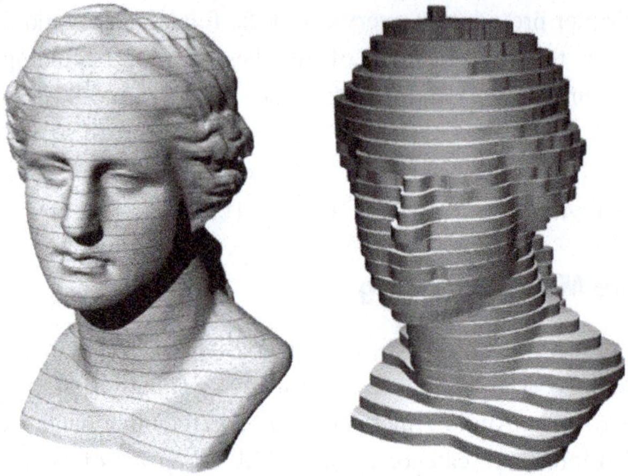

Figure 2-2: Principle of AM - manufacturing and merging of single layers to a 3D object [65].

Wegener et al. [233] listed more than 100 different AM technologies, and several are being developed every year. They differ, among other things, in the material they can process, the physical principle of how the material is "bonded," the size of the parts that can be printed, and the complexity of the machines. The most common materials are plastics, metals and ceramics. Furthermore, multi-material printing is becoming more and more important [240]. Due to the variety of different AM processes, there are several ways to categorize them, e.g., by binding mechanism, energy source, or the aggregate state of the material (gas, liquid or solid). Figure 2-3 provides an overview of the most relevant industrial AM technologies, based on their build rate, resolution and investment costs. Selective Laser Melting (SLM) is the most relevant AM technology for metal parts, comprising a market share of more than 70% [10, 240].

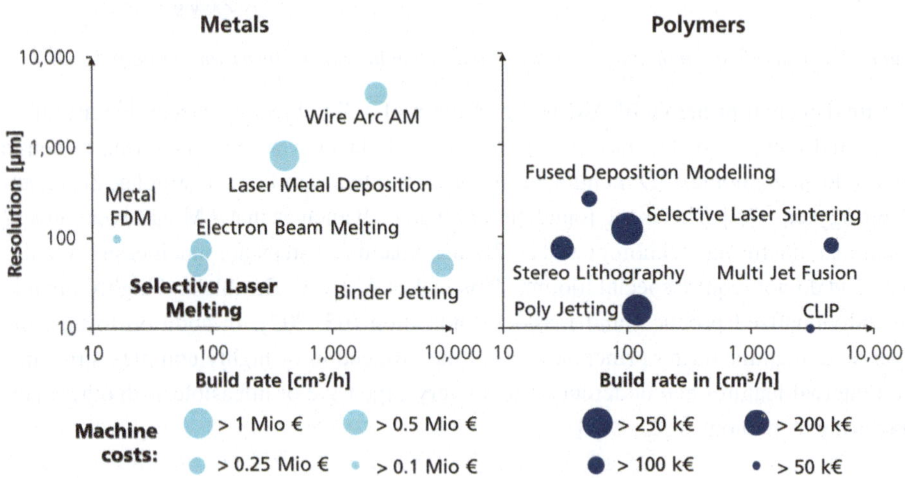

Figure 2-3: Most relevant industrial AM technologies based on build rate, resolution and investment costs [50].

2.2 Selective Laser Melting

The present thesis investigates the effect of the laser beam profile on the *Selective Laser Melting* (*SLM*) process. Other common names for the technology are *Laser Beam Melting (LBM)*, *Laser Powder Bed Fusion (LPBF)*, *Laser Metal Fusion (LMF)*, *Laser Cusing* or *Direct Metal Laser Sintering (DMLS)*. The DIN EN ISO / ASTM 52900:2018-06 2018 labels the technology as a *Powder Bed Fusion technology [45]*. The process was mainly developed by Dr. Fockele and Dr. Schwarze in the mid-1990s, in cooperation with the Fraunhofer ILT, and has been commercialized by several companies since [22]. The machine manufacturers with the largest market share are EOS Optical Systems GmbH, Concept Laser GmbH und SLM Solutions Group AG, all located in Germany [240].

The following chapters provide an overview of the fundamentals of SLM that are essential to understand the thesis. In the beginning, an introduction into the SLM process (Chapter 2.2.1), the process chain (Chapter 2.2.2), as well as advantages and challenges (Chapter 2.2.3) are given. Furthermore, the main process parameters (Chapter 2.2.4), the details of the laser-material interaction as well as the melt pool dynamics (Chapter 2.2.5) are explained and the formation of the relevant defects is described (Chapter 2.2.6).

2.2.1 General Principle

The general principle of the SLM process and an illustration of the core components of an SLM machine are shown in Figure 2-4 and Figure 2-5.

Before the actual build process starts, the 3D-CAD (Computer Aided Design) model is sliced into layers of desired thickness and transferred to the machine [4]. The build process consists of three consecutive process steps that are repeated every single layer [149, 164]:

- *Application of the powder layer*: The powder material is supplied from a container or reservoir and a thin powder-layer is applied with the help of a recoating device. Typical layer thickness varies from 30-100 µm.
- *Exposure*: A focused laser beam melts the metallic powder according to the part's geometric data in the specific layer. The molten powder fuses with the solid layer underneath, creating a metallurgical bonding. The beam is positioned in x- and y-directions with galvanometer scanners that are located above the build chamber. A gas stream right above the build platform ensures that metal vapor and spatters are blown away from the interaction of the laser zone with the material.
- *Lowering of build platform*: The build platform is lowered by the thickness of one layer.

The iterative process is repeated until the whole part is built [4]. Afterwards, the part can be removed from the machine.

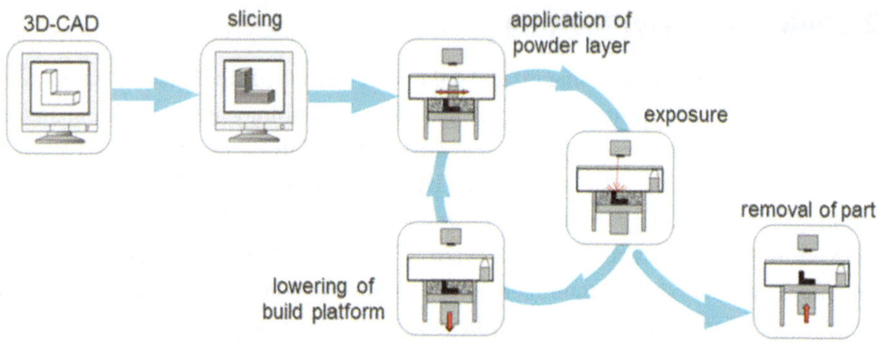

Figure 2-4: General principle of the SLM process (based on [228]).

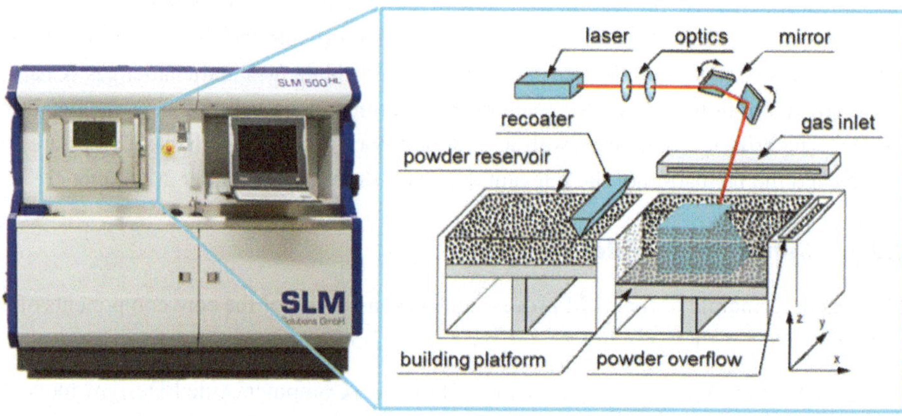

Figure 2-5: Illustration of an SLM processing chamber with main components (based on [22]).

A variety of different materials can be processed by SLM. The most common materials are aluminum, titanium, nickel-base and steel alloys [243, 252]. Due to the complete melting of the powder particles during the process, a residual porosity of < 0.5% can be achieved (see Figure 2-6-a) [223]. This, combined with a fine microstructure, results in excellent mechanical properties [22, 164, 242]. However, the quality of the surface of SLM parts is limited by powder particle size and shape (see Figure 2-6-b) and is comparable to the surface of casted parts. The typical surface roughness varies between R_a = 16-45 µm and depends on the surface's orientation to the build direction [65, 242]. Also, support-structures are necessary to fix the part and guarantee a sufficient heat transfer at the location of laser-material interaction when the surface angle is lower than approximately 45° to build direction (see Figure 2-6-c) [22, 65].

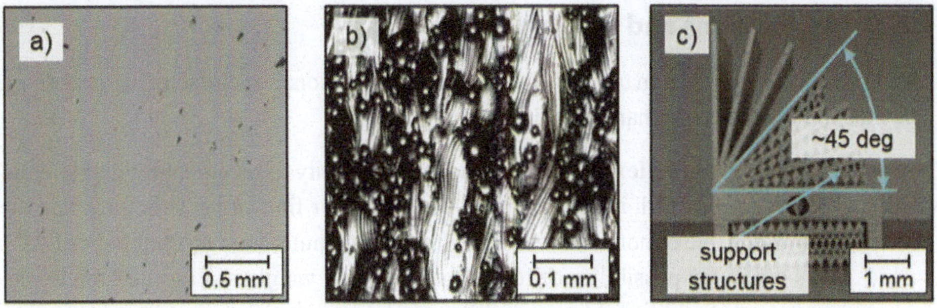

Figure 2-6: a) Micro-section of SLM part with low porosity; b) Scanning electron microscope (SEM) picture of SLM surface showing powder adhesion; c) support-structures of SLM part.

2.2.2 Overview of Process Chain

In addition to the actual build process, supplementary processing steps are necessary to manufacture a final part. The processing steps can be divided into pre-process, in-process and post-process steps, as shown in Figure 2-7 [229].

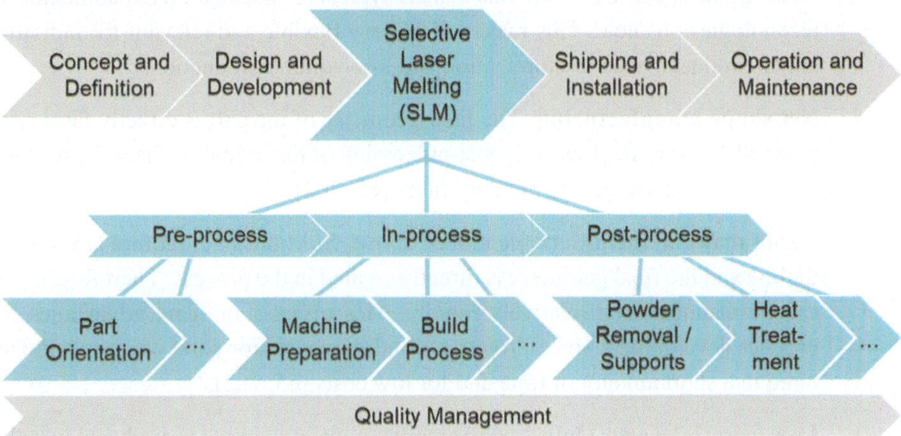

Figure 2-7: SLM process chain (based on [22]).

Pre-process steps: In this category, all operations are compiled that take place before the SLM process starts, such as the definition of support structures and generation of machine-specific data.

In-process steps: In this category, all operations that are necessary to operate the SLM machine are compiled, such as powder sieving and preparation of the build platform.

Post-process steps: In this category, all operations that take place after the SLM process are compiled, such as removing the support structures and the cleaning of the part. Optional steps such as heat treatment or surface finishing are usually performed to optimize the part quality.

2.2.3 Advantages and Challenges

The SLM technology has many advantages over conventional manufacturing processes. The most relevant are summarized below [22, 65, 202]:

- **Freedom of complexity**: SLM offers the possibility to manufacture highly complex structures with internal features, undercuts or fine lattice structures that are very expensive or not at all feasible with other manufacturing technologies [202]. This offers the possibility of realizing significant weight savings and the implementation of additional functionalities, such as a plagiarism protection or a compressed air ejector for injection molding [102, 103]. Another advantage is that the complexity "comes along for free." This means that the manufacturing costs are not significantly influenced by the complexity of a part, making the technology particularly interesting for complex structures [65, 202].

- **Tool-less manufacturing**: The manufacturing process does not require any special tooling [22, 202]. The part can be created directly out of the 3D CAD data, making part-specific investments that are typical for conventional manufacturing technologies obsolete. This makes SLM very flexible and relevant for individualized products, mass-customization and low-volume production.

- **Net-shape manufacturing**: The final geometry of the part is directly fabricated in the SLM process. However, post-processing of functional surfaces, by milling, for example, is mandatory in many cases [65, 202].

- **Rapid manufacturing**: As mentioned above, SLM does not require any special tooling and the final geometry is directly created in the process. Therefore, a part can be manufactured relatively rapidly. This offers significant advantages for product development since prototypes and demonstrator parts can be manufactured in a short amount of time and for low costs [65, 202].

- **Low material waste**: Only a small amount of material waste is produced (usually $< 5\%$), as the powder that is not molten, can be reused [22, 144]. This makes SLM relevant for parts where a high amount of material has to be removed in subtractive processes.

- **Good mechanical properties**: Highly dense parts with a fine microstructure can be created with SLM. This results in excellent mechanical properties that are, in some aspects, superior to the properties that are achieved by conventional processes [202, 242].

- **Suitability for Industry 4.0**: SLM is suitable for realizing highly flexible, integrated and automatized production, which leverages the potential of *Industry 4.0*. The main reason for this is its high flexibility and the digital nature of the process, making it easy to automatize [202, 227].

Despite the advantages, SLM faces challenges, the most significant of which are summarized below [22, 188, 202]:

- **Low productivity and high part costs**: The build-up speed of SLM machines varies from between 10-100 cm^3/h, depending on the type of machine, the material used and the process parameters [188, 240]. This can lead to manufacturing times of several days for large parts. Since the most significant cost-driver in SLM is the machine-hour rate, high part costs are the result. Therefore, a significant increase in productivity has to be realized to make SLM cost-competitive for medium to high-volume production [22, 188].

- **Necessity of support structures**: Support structures are necessary to fixate the part on the build platform, ensure heat conduction away from the laser interaction zone into the base plate, and counteract the residual stresses (see Figure 2-6-c) [185]. The support structures are removed in an additional post-processing step, usually manually [202, 228].

- **Residual stresses**: The local heat input of the laser beam can provoke significant residual stresses within the part [164, 181]. This can lead to distortion or failure of the part when the support-structures are not appropriately designed. In most cases, the residual stresses are not desired and need to be removed with a subsequent heat treatment, before the part can be detached from the build platform [22].

- **Low surface quality**: The surface quality in the SLM process is limited by the powder material, which adheres to the surface or is only partially molten (see Figure 2-6-b) [22, 202]. The comparably rough surface harms the mechanical properties, especially concerning fatigue [242]. As a consequence, a surface finishing might be necessary to improve the surface quality [22].

- **Limited reproducibility**: SLM is a complex process with over 130 parameters that influence the final part quality [184]. Not all of the influences are fully understood and the control of some critical parameters has not been realized yet. This leads to process instabilities that result in part deformation, increased porosity or cracks [181, 244]. Therefore, significant experience with the operation of an SLM machine is necessary to achieve a stable process.

- **Quality assurance**: The lack of reproducibility makes suitable quality assurance essential. In many cases, subsequent quality assurance steps after the build job are necessary, e.g., by Computer Tomography (CT) or density measurements [22, 240, 243]. This can increase the part costs and lead time significantly. However, AM's layer-wise manufacturing principle provides the unique possibility to monitor the manufacturing process in high detail, for example by measuring the melt pool emissions "on-line" with a photo-diode. Therefore, on-line monitoring can provide an efficient solution to directly measure the part-quality during the build process in the future [52].

- **Low degree of industrialization**: Although the digital nature of the technology enables full automation of the process, this potential has not been exploited yet. State of the art SLM machines require special data preparation and do not use standardized file formats [202, 227]. This prevents lean integration into the process chain of conventional production processes. On top of this, there is still a significant amount of manual work necessary to operate an SLM machine, especially for machine preparation and part removal after the build process [22].

2.2.4 Essential Process Parameters and their Influence

Figure 2-8 shows the essential SLM process parameters. They have a strong interdependency, which makes it necessary to match them in order to achieve a stable process and highly dense parts [22, 76]. The optimization of process parameters is usually done experimentally, but simulations can reduce the amount of experimental work [9]. This chapter gives an overview of the influence of the most significant process parameters. A detailed description of the melting process and its important phenomena is given in Chapter 2.2.5.

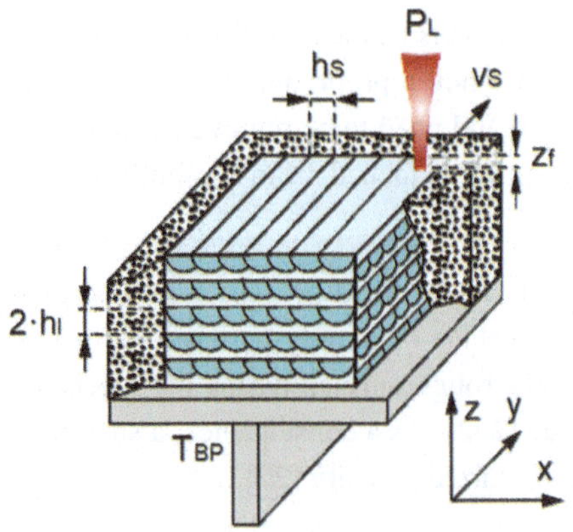

P_L:	Laser power [W]
v_s:	Scanning speed [mm/s]
h_s:	Hatch distance [µm]
h_l:	Leveling height of base plate [µm]
T_{BP}:	Preheating temp. of base plate [K]
d_f:	Focus diameter of laser beam [µm]
z_f:	Focus height [mm]

Figure 2-8: Essential process parameters of SLM [22, 149, 164].

Laser Power P_L, Scanning Speed v_s and Linear Energy Density (LED)

The laser power P_L and the scanning speed v_s have a significant effect on the process stability and the porosity of the final part [22, 76]. The ratio of laser power P_L to the scanning speed v_S is defined as the *linear energy density (LED)* as shown in Equation 2.1 [22, 149].

$$LED = P_L/v_S \tag{2.1}$$

LED: Linear energy density [J/mm] P_L: Laser power [W]
v_S: Scanning speed [mm/s]

It is essential to melt the powder completely to manufacture highly dense parts and create a metallurgical bonding with the subsequent layer (see Figure 2-9) [149]. When the LED is too low, the powder particles are only partially molten and discontinuous melt tracks as well as increased porosity are the result [1, 30, 105, 247]. With an increase in LED, the melt pool width and depth are increased and continuous melt tracks are formed [16, 247, 254]. Gong et al. [69] found a linear increase in melt pool width for single tracks for TiAl6V4 with an increase in laser power at constant scanning speed. A negative effect of an increase in LED is that vaporization and spatters are intensified, leading to porosity [139, 149, 170, 210]. Additionally, the viscosity of the liquefied metal is reduced at higher temperatures. This increases the importance of hydrodynamic effects, such as Marangoni flow, leading to instabilities in the process (for more details, see Chapter 2.2.5) [22, 69, 76, 98]. A high LED can also provoke keyhole porosity [1, 69, 139, 221], especially at the end of scan tracks [141]. The application of a pulsed laser can reduce some of the adverse effects [76]. Furthermore, higher LEDs lead to a coarser microstructure [22, 76] and fewer residual stresses [170].

It has to be noted that the LED alone is not a sufficient criterion for a stable SLM process. At a constant LED, the process becomes unstable for high values of laser power and scanning speed due to overheating and increased vaporization [30, 256].

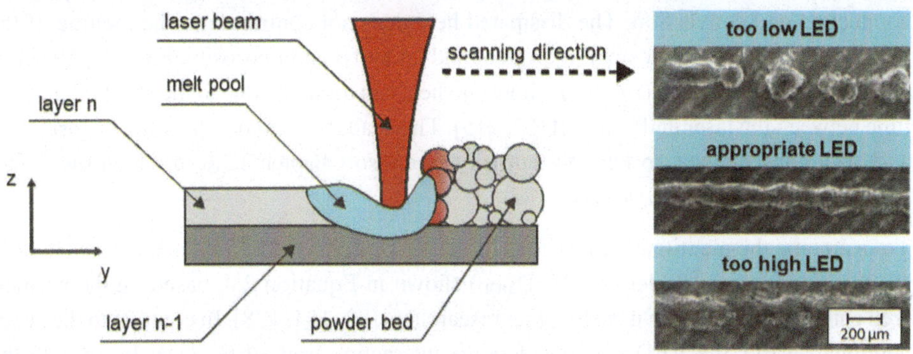

Figure 2-9: Side view of the melting process in SLM and images of single melt tracks with too low, correct and too high LED (sketch based on [164]; images from [96]).

Hatch Distance h_s and Surface Energy Density (SED)

The hatch distance h_s defines the distance between two adjacent melt tracks (see Figure 2-8). The LED divided by the hatch distance hs defines the *surface energy density (SED)*, as shown in Equation 2.2.

$$SED = P_L/(v_S \cdot h_S) \quad (2.2)$$

SED:	Surface energy density [J/mm^2]	P_L:	Laser power [W]
v_S:	Scanning speed [mm/s]	h_S:	Hatch distance [µm]

Several researchers have shown that an overlap of ~30% between adjacent melt tracks provides high densities [149, 164, 170]. A hatch distance that is too high leads to inter- and intra-layer porosity [1] and a hatch distance that is too low leads to irregular melt pool shapes, increased porosity [149, 223] and low productivity.

Leveling height of build platform h_l, Volumetric Energy Density (VED) and modified Volumetric Energy Density (VED_{MOD})

The leveling height h_l defines the distance the build platform is leveled down after each layer. The SED, divided by the leveling height, defines the *volumetric energy density (VED)* as shown in Equation 2.3. Many researchers use the VED (see Equation 2.3) as a design parameter for the process and to describe parameters that lead to a stable process and dense parts [22, 92]. However, although the VED has shown that it can accurately predict the minimum energy input that is necessary to avoid lack-of-fusion porosity (see Chapter 2.2.6), it must be noted that the VED is not a holistic design parameter for the process [16, 22, 54]. The main reason for this is that the VED does not consider the interaction time between the laser and the material, as it is constant for a uniform ratio of laser power P_L to scanning speed v_S and also does not consider essential material properties [54, 154, 205]. As a result, the same VED can lead to lack-of-fusion porosity, a stable process or keyhole porosity, depending on the actual parameters used as schematically shown in Figure 2-10. For low laser power and scanning speed ($x_1 << 1$), the interaction time between the laser and material is long and a significant amount of the heat can dissipate via conduction and convection. The dissipated heat does not contribute to the melting of the material, which leads to a small melt pool and lack-of-fusion porosity. For a high energy input and scanning speed ($x_2 >> 1$), almost no heat can dissipate due to the short interaction time between the laser and material [54, 215]. This leads to local overheating, vaporization and, as a consequence, to keyhole porosity (see also Chapter 2.2.6 in which the defect generation is described in detail).

To overcome the mentioned limitations of the VED, Ferro et al. [54] have defined a *modified* volumetric energy density (VED_{MOD}) shown in Equation 2.4, based on the normalized enthalpy formulation used by other researchers [92, 154, 178]. In contrast to the "conventional" VED, the VED_{MOD} considers the interaction time of the laser beam with the material and essential material properties. Both, the normalized enthalpy formula and VED_{MOD} have demonstrated to predict the process parameters for which keyhole porosity occurs more accurately than the conventional VED [54, 154, 178].

$$VED = P_L/(v_S \cdot h_S \cdot h_l) \tag{2.3}$$

$$VED_{MOD} = \frac{a_{fres}}{h_S}\sqrt{\frac{C_p \cdot \rho}{4 \cdot k \cdot d_f}} \cdot \frac{P_L}{\sqrt{v_S}} \tag{2.4}$$

VED:	Volumetric energy density [J/mm³]		
VED_{MOD}:	*Modified* volumetric energy density [J/mm³]		
P_L:	Laser Power [W]	v_S:	Scanning speed [mm/s]
h_S:	Hatch distance [µm]	h_l:	Leveling height of build platform [µm]
a_{fres}:	Absorption coefficient [1]	C_p:	Specific heat capacity [J/kgK]
ρ:	Density of material [kg/m³]	d_f:	Focus diameter of laser beam [µm]
k:	Thermal conductivity [W/(m·K)]		

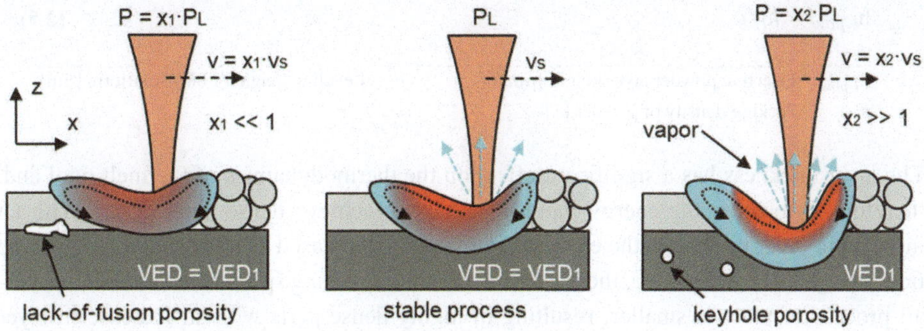

Figure 2-10: Formation of defects for different combinations of laser power P_L and scanning speed v_S at constant VED.

It is important to notice that the thickness of the powder layer on top of a previous molten layer is higher than the layer thickness of the molten material, as shown in Figure 2-11. The powder layer height increases in the first layers due to power fall in and densification during melting, until it reaches its theoretical value of h_{l_pow} after approximately seven layers (see Equation 2.5) [149, 164]. When applying this equation, it has to be noted that the packing density in a single layer is lower than the value measured by standard powder characterization methods, as shown in Figure 2-12 [156, 238]. Experimental investigations have shown that the powder layer height is further increased due to powder depletion during the process, such as spatters, vaporization and denudation. This can result in a powder layer height that is more than four times the leveling height of the build platform [156, 238].

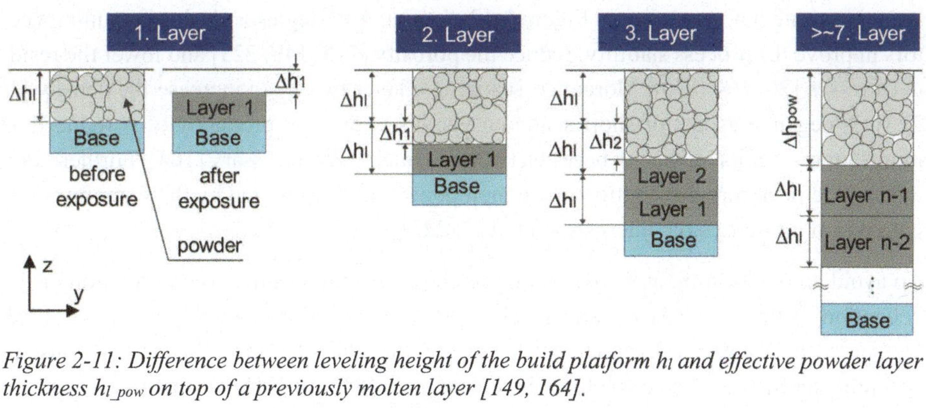

Figure 2-11: Difference between leveling height of the build platform h_l and effective powder layer thickness h_{l_pow} on top of a previously molten layer [149, 164].

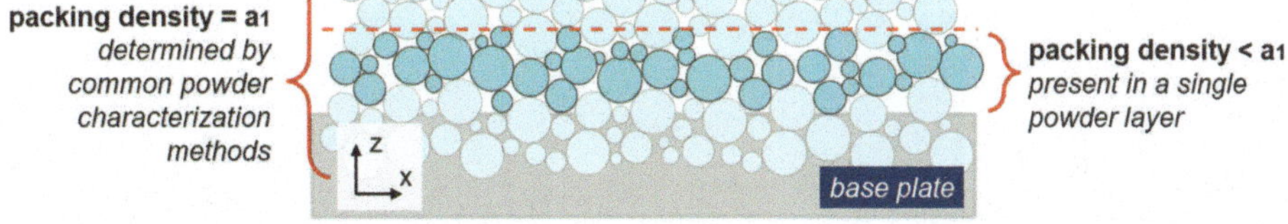

Figure 2-12: Reduced packing density of a single powder layer in SLM [238].

$$\mathrm{h}_{l\_pow} = h_l / a \tag{2.5}$$

h_{l_pow}: Effective powder layer height [µm] h_l: Leveling height of build platform [µm]
a: Packing density of powder [1]

The layer thickness has a significant effect on the thermodynamics of the melt pool and, therefore, influences the energy input necessary to achieve dense parts [232]. With an increase in layer thickness, the energy input must be increased as more material has to be liquefied [170]. Furthermore, the maximum feasible scanning speed is lower and the overall process window is smaller, resulting in highly dense parts with an increase in layer thickness [170]. Despite the reduction in scanning speed, the overall productivity of the process can be increased, as more material is molten for every single scan track and due to the circumstance that the total number of layers is reduced [149]. Furthermore, the microstructure and mechanical properties are influenced [218], the residual stresses can be lowered [164, 170] and the accuracy of the part is aggravated due to the staircase effect by increasing the layer thickness [65, 149].

Pre-heating Temperature T_{BP}

Some SLM machines are equipped with a resistive heating system in the base plate. Preheating decreases the VED that is necessary to melt the material completely. As a result, higher scanning speeds can be used [170]. In addition, the lower temperature gradient reduces the residual stresses in the part [164, 170] and influences the microstructure [170].

Scanning Strategy

The scanning strategy defines the sequence in which the laser scans over a layer. Typical scanning strategies are shown in Figure 2-13. Scanning strategies with short scanning vectors improve the process stability, reduce the porosity [113, 149, 223] and lower the residual stresses [76, 164, 170]. Moreover, scanning a layer twice can increase the density [1, 223]. A negative aspect of short scanning vectors is that the productivity is reduced as more "jump-vectors" (the path between two scan tracks) are necessary [164]. Furthermore, the local temperature distribution in the melt pool vicinity is altered by the scanning strategy, which influences the microstructure [22, 223].

To avoid an overheating at the beginning or end of scan tracks due to acceleration or deceleration of the laser, laser scanner delays or sky-writing are typically used [81, 177]. Compared to using laser scanner delays, sky-writing has the advantage that a constant scanning speed is achieved for the whole scan path [81, 138]. However, it also leads to longer scan paths and, consequently, to a decrease in productivity [138, 146].

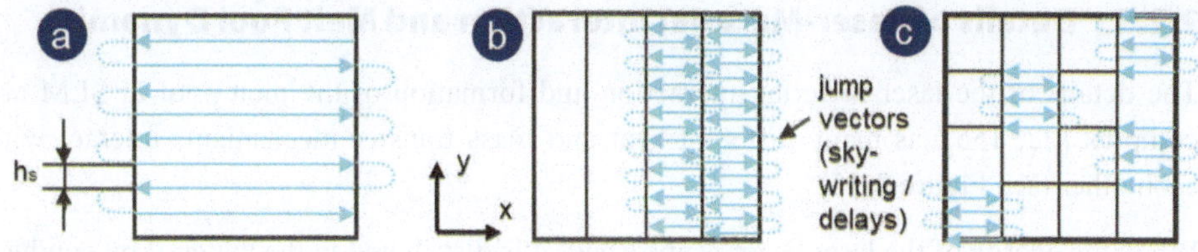

Figure 2-13: Typical scanning strategies for SLM; a) continuous; b) stripe; c) checker-board (based on [164]).

Beam Diameter d_f and Focus Position z_f

By moving the laser focus position above or underneath the top of the working plane, the diameter of the laser beam in the working plane can be changed (see Figure 2-14) [49]. By increasing the beam diameter, the overall productivity can be increased based on a larger hatch distance and higher laser powers that can be used [170]. This is utilized in the "hull-core" strategy, where the part is divided into two sections, as shown in Figure 2-14 [22]. For this, the hull is processed with a small beam diameter to guarantee a high resolution at the surface and the core, where a high resolution is not essential, is processed with a much larger beam diameter and higher build rates. The productivity can be further increased when the core is processed with a multiple of the layer thickness of the hull. Moreover, the beam diameter influences the cooling-rate of the melt pool and thus, the microstructure. Schwarze et al. produced large, highly oriented grains for 316L in the core with a beam diameter of ~700 µm [204].

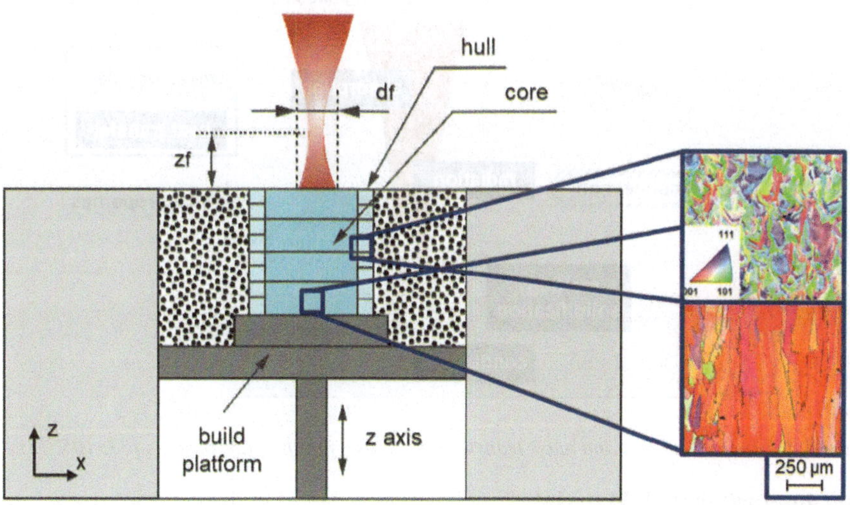

Figure 2-14: Concept of the hull-core strategy and variation of microstructure for 316L stainless steel (images of microstructures taken from [204]).

2.2.5 Details of Laser-Material Interaction and Melt Pool Dynamics

The details of the laser-material interaction and formation of the melt pool in SLM are complex [22, 155], as many different heat and mass transfer mechanisms interact with each other (see Figure 2-15).

When the energy of the laser beam is absorbed, it is distributed in the material by conduction and convection [22, 132, 203]. Several studies have shown that especially the capillary forces, Marangoni-effect (thermo-capillary-force), recoil pressure due to vaporization and conduction in the solidified material are essential for the dynamics of the process [11, 22, 143, 181]. The heat conduction between single powder particles is very low (between 1-15% of the corresponding bulk material) and has a minor influence [42, 132]. The same applies to buoyancy forces [97, 101, 139] and radiation [98, 127]. The process is also influenced by the characteristics of the powder bed and its randomness [22, 132]. The high dynamic of the process makes it difficult to investigate the details of the laser-material interaction and the melt pool dynamics experimentally [76, 164]. However, simulations [97, 141], high-speed videos [143, 165] and x-ray diffraction imaging have improved the understanding of the process significantly in recent years [221, 261].

In the following, a more detailed description of the laser interaction with the material as well as the effect of (thermo-)capillary forces and vaporization on the melt pool characteristics is provided.

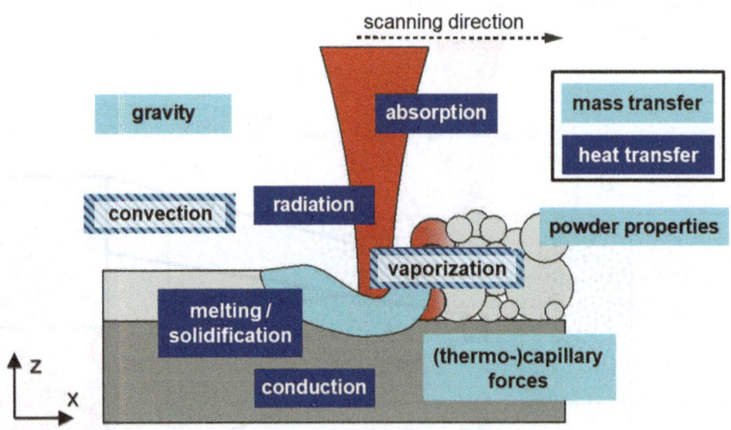

Figure 2-15: Important heat and mass transfer mechanisms of the SLM process [37, 105, 139].

Laser Interaction with Material

In SLM, the energy is deposited into the material by a moving laser beam. The laser beam irradiates the top surface of the powder and melt pool, where it is partially absorbed and reflected. Due to the higher surface area of the powder and multiple reflections, the absorption of the powder bed is significantly higher than the one of solid bulk material [97, 101, 149, 252]. King et al. [101] measured an absorptivity of 55-65% for SS316L, TiAl6V4 and pure aluminum powders while the absorptivity of the corresponding bulk material varies between 5-40% [7, 21, 216]. It is important to notice that the powder particles melt at the front edge of the laser beam [143] and that the center of the melt pool is slightly "behind" the center of the laser beam, as illustrated in Figure 2-16 [203]. Therefore, most of the energy is absorbed by the molten material and the absorptivity of the liquefied metal has a predominant effect on the total energy balance [98]. A high amount of absorption is desirable, as it has a positive effect on the stability and energy efficiency of the process [22, 252].

There are many influences on the absorptivity of a laser beam, such as the wavelength, intensity and polarization of the laser beam, the angle of incidence and chemical composition of the surface (e.g., oxide layers) [22, 97, 191, 210]. Furthermore, the absorptivity increases with temperature and, therefore, changes dynamically during the process [22]. In addition to the physiochemical properties, the morphology and randomness of the powder bed, the surface and shape of the particles, and the melt pool topology also affect absorptivity [97, 106, 149, 247]. Additionally, plasma and metal vapor generated due to vaporization can interfere with the laser beam and reduce the absorption [98]. Moreover, multiple reflections in the depression zone around the laser beam center can increase the absorptivity (see Figure 2-16). Zielinski et al. [267] measured an absorptivity during the SLM process and determined values between 50–80% for TiAl6V4, which is in good agreement with values obtained by Matthews at al. for Cu, W, TiAl6V4, AlSl 316L and Inconel 625 [190].

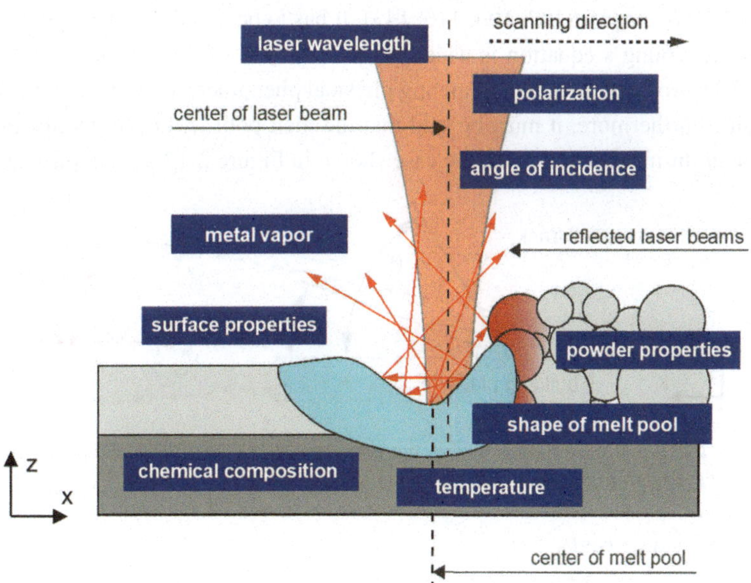

Figure 2-16: Laser interaction with the melt pool and influence parameters on the absorptivity.

Influence of (Thermo-)Capillary Forces and Vaporization on the Melt Pool

For a stable SLM process, a single melt track with a shape of a circle or segmental cylinder, as shown in Figure 2-17 is desired [247], with a width that is usually two to three times the diameter of the laser beam [143]. The final shape of the solidified melt track is a result of several interrelated physical phenomena acting on the melt pool. Most relevant are the capillary forces (surface tension), the Marangoni effect (thermo-capillary forces) as well as the recoil pressure due to vaporization [132, 139, 143, 181]. In contrast to this, the buoyancy forces and density-driven convective flow have a minor effect [97, 98, 110].

The influence of capillary forces, thermo-capillary forces and vapor pressure will be explained in more detail in the following paragraphs.

Influence of Capillary Forces on Heat and Mass Transfer in SLM
Capillary forces, surface tension and wettability are highly related phenomena governed by surface and interface energies [14, 105, 128]. For a liquid on a solid material, the total surface energy is at its minimum when the liquid forms a smooth surface and the wetting angle θ equals the equilibrium wetting angle θ_0, as defined by Young's equation shown in Equation 2.6 [76, 109, 139, 231]. For all other cases, the capillary force F_{cap} (see Equation 2.7) will act on the solid-liquid-gas-interface to bring the interface into the equilibrium state [53, 76, 105]. During SLM, the capillary force F_{cap} "pulls" the melt pool sideways when the liquefied metal comes into contact with the solid layer underneath, increasing the melt pool width and heat transfer into the base plate [97]. Additionally, wetting forces can drag surrounding powder particles into the melt pool when the molten material gets in contact with the surface of a loose powder particle [143]. This increases the melt pool size and contributes to the denudation zone observed around solidified melt tracks, as shown in Figure 2-20 on page 25 [97, 106, 110, 143]. It has to be noted that the equilibrium state described by Young's equation is usually not reached during SLM, due to the high dynamics of the process and other competing physical phenomena described in the following paragraphs. Furthermore, it must be said that the melt pool will penetrate the previously molten layer during SLM and that the case shown in Figure 2-17 is a simplification.

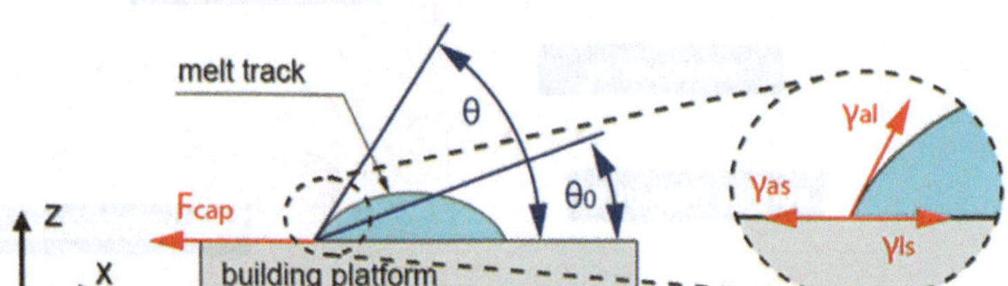

Figure 2-17: Capillary force F_{cap} as well as equilibrium (θ_0) and dynamic wetting angle (θ) between the substrate, fluid and ambient gas (based on [105, 189, 231]).

$$\gamma_{as} - \gamma_{ls} = \gamma_{al} \cdot \cos\theta_0 \qquad (2.6)$$

$$F_{cap} = \gamma_{as} - \gamma_{ls} - \gamma_{al} \cdot \cos\theta \qquad (2.7)$$

γ_{as}: Interf. tension at air-solid interface [N/m] γ_{ls}: Interf. tension at liquid-solid interface [N/m]
γ_{al}: Interf. tension at air-liquid interface [N/m] θ_0: Equilibrium wetting angle [rad; °]
θ: Dynamic wetting angle [rad; °] F_{cap}: Capillary force [N/m]

One common cause for defects that is highly governed by the capillary forces is the *balling effect* [72, 97, 189]. Balling describes the break-up of a continuous melt track into several single segments, as shown in Figure 2-18 and is related to the Plateau-Rayleigh instability [98, 139, 247]. It occurs when the free surface energy of a liquid is smaller for a sphere than for a cylindrical shape. For a free cylindrical fluid, balling will appear when the length-to-width ratio is larger than $1.2 \cdot \pi$, as shown in Equation 2.8 (Plateau-Rayleigh instability) [110, 189, 247]. While this assumption is in line with experiments for single tracks in a loose powder bed [29], different conditions apply for the melting on a solid substrate. For a segment of a cylinder on a flat solid surface and a wetting angle of $\theta > \pi/2$, the cylindrical shape is stable when the condition in Equation 2.9 applies (see Figure 2-18) [247]. For a wetting angle of $\theta < \pi/2$ the cylindrical shape is always stable [189, 247]. It must be mentioned that Equation 2.9 neglects the fact that the melt pool penetrates the base plate during SLM as well as other phenomena that affect the balling tendency as the melt flow or vaporization [99]. This can explain experimental results in which balling has been reported for $l_{mp}/w_{mp} > 2.1$ [22, 109, 110].

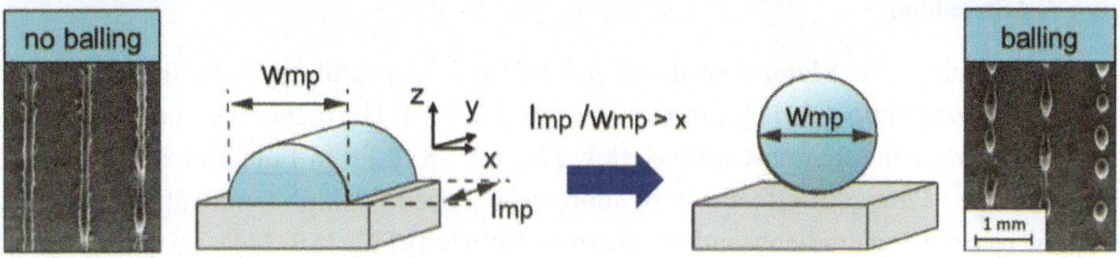

Figure 2-18: Visualization of the balling effect [22] (images of single tracks from [128]).

$$\frac{l_{mp}}{w_{mp}} < \sqrt{\frac{3}{2}}\pi \tag{2.8}$$

$$\frac{l_{mp}}{w_{mp}} < \frac{\pi}{\sqrt{2}}\sqrt{\frac{2\theta(2+\cos 2\theta) - 3\sin 2\theta}{\theta(1+\cos 2\theta) - \sin 2\theta}} \tag{2.9}$$

l_{mp}: Length of the cylindrical segment [mm] w_{mp}: Width of the cylindrical segment [mm]
θ: Dynamic wetting angle [rad; °]

Several material and process characteristics influence the balling tendency [72, 128]. In general, a wide and short melt pool is favorable to avoid balling (see Equations 2.8 and 2.9), which can be achieved by low scanning speeds [72, 97, 128]. Furthermore, it has been reported that re-melting and high packing densities of the powder layer reduce the balling tendency [98, 120, 128]. In contrast to that, high scan speeds create a long and thin melt pool, which reinforces balling [72, 110, 247]. Another factor that increases the balling tendency is oxidation, as it deteriorates the wetting characteristics of the fluid [110, 128, 189].

Influence of Thermo-Capillary Flow on Heat and Mass Transfer in SLM

The Marangoni effect (or thermo-capillary effect) results from the temperature dependence of surface tension [11, 97, 189]. When a gradient in surface tension exists in a liquid, a surface driven flow from the area with low surface tension to the area with high surface tension occurs [22]. For most metals, the surface tension decreases with temperature, leading to an "outward" flow of the material, as shown in Figure 2-19 [98, 136, 189, 254]. This creates a shallow melt pool and amplifies the heat dissipation as well as melt pool width [22, 29, 254]. The underlying temperature gradient results from the imbalanced heat input from the Gaussian laser beam profile and surrounding material acting as a heat sink [22, 98, 244]. However, the surface tension and its temperature coefficient are affected by surface-active elements, such as oxygen or sulfur [62, 110, 136, 253]. With an increase of surface-active elements, the surface tension is reduced and its temperature coefficient increased. At a certain threshold, the surface tension coefficient changes sign resulting in a surface tension that increases with temperature [62, 110, 193, 262]. At this threshold, the direction of Marangoni flow is inverted and a narrow and deep melt pool is the result (see Figure 2-19) [62]. The effect of surface-active elements on the SLM process is not well understood [62], but insight from other processes, such as Laser Metal Deposition [62], welding [193, 263] or Laser Surface Melting [122] can be transferred to the SLM process to a certain extend.

The magnitude of the Marangoni flow can be described quantitatively by the dimensionless *Marangoni number Ma* (see Equation 2.10) [11, 189]. The higher the Marangoni number, the stronger the thermo-capillary flow [11, 62, 189]. From Equation 2.10, it can be seen that the Marangoni effect will be more pronounced for high temperature gradients, high surface tension gradients and low thermal diffusivity [22, 189, 244]. To quantitatively describe the contribution of the Marangoni flow to the overall heat transfer, the *Peclet number Pe* can be used (see Equation 2.11) [62, 182]. The Peclet number is defined as the ratio of heat transfer by convection to heat transfer by conduction [130, 194, 251]. Peclet numbers up to a value of 100 have been reported for SLM [62], which demonstrates the significance of convective heat transport for the process. Although the increased heat transfer due to Marangoni flow positively affects the SLM process, it can cause instabilities when the melt flow becomes too strong [72, 98, 244].

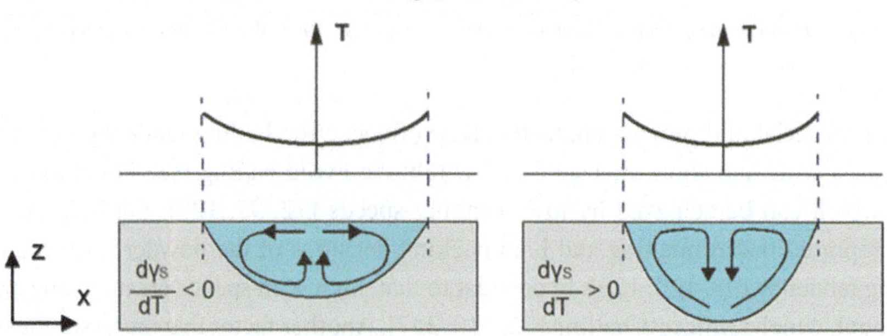

Figure 2-19: Effect of temperature-dependent surface tension on Marangoni flow and shape of the melt pool [22, 42].

$$Ma = \frac{d\gamma_s}{dT}\frac{dT}{dx}\frac{l_{c\_mp}}{2\mu\delta} = \frac{d\gamma_s}{dT}\frac{dT}{dx}\frac{l_{c\_mp}\rho C_p}{2\mu k} \tag{2.10}$$

$$Pe = \frac{heat\ transfer\ by\ convection}{heat\ transfer\ by\ conduction} = \frac{w_l \cdot v_m \cdot \rho \cdot C_p}{k} \tag{2.11}$$

Ma:	Marangoni number [1]	$\frac{d\gamma_s}{dT}$:	Surface tension gradient [N/(m·K)]
$\frac{dT}{dx}$:	Temperature gradient [K/m]	l_{c_mp}:	Characteristic length (width of melt pool) [m]
μ:	Viscosity of fluid [N·s/m²]	δ:	Thermal diffusivity [m²/s]
Pe:	Peclet number [1]	v_m	Velocity of melt flow [m/s]
ρ:	Density [g/cm³]	k:	Thermal conductivity [W/(m·K)]
C_p:	Specific heat capacity [J/(kg·K)]		

Effect of Recoil Pressure on Heat and Mass Transfer in SLM

During the melting process, the melt pool surface can reach boiling temperature due to the high intensity of the laser beam. As a result, a vapor plume is ejected from the melt pool that induces a recoil pressure on the melt pool surfaces shown in Figure 2-20, which can interfere with the laser beam [132, 189]. Although the loss of material due to vaporization is small (usually < 1%) [98, 101], a significant amount of energy is lost due to the high enthalpy of vaporization [98, 101]. Moreover, the chemical composition of the material can be changed [139, 189, 210]. Furthermore, the recoil pressure creates a depression zone underneath the laser beam, which influences the heat and mass transfer in the melt pool [101, 143, 261]. The recoil pressure increases with temperature [132, 143] and can increase the melt pool velocities to several meters per second [137]. Additionally, the vapor induces shear stresses on the melt pool surface that adds a surface-driven flow similar to the Marangoni flow [150].

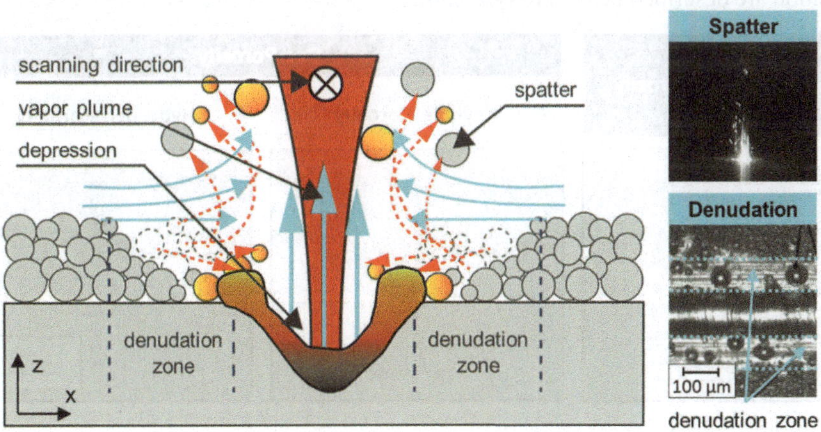

Figure 2-20: Depression and denudation zone as well as spatters in SLM (images of spatters and denudation zone are from [149, 247]).

Vaporization is also a major cause of the spatters that are generated during the melting process (see Figure 2-20). Spatters are an inadvertent phenomenon as they lead to instabilities and can cause defects when they fall onto the powder bed [261]. Spatters can be ejected directly from the melt pool [98, 101, 149, 261] when the radial pressure gradient (as a result of the vaporization) and the kinetic energy of liquid overcome the surface tension forces [189, 224]. Additionally, spatters are caused when surrounding particles are entrained, swept-up and sideways due to the gas flow as shown in Figure 2-20. This gas flow results from the pressure drop induced by the vapor plum as described by the *Bernoulli Effect* [22, 132, 137]. The entrainment of surrounding powder particles is one of the phenomena contributing to the denudation zone [143, 261] that is defined as the area in the vicinity of a melt track with a reduced amount of powder. Other phenomena that contribute to the denudation zone are the entrainment of powder particles into the melt pool by wetting forces [98, 137, 246, 261], and the sideway propelling of particles when they are partially heated by the laser beam [143]. The latter can also push away additional powder particles due to momentum transfer, increasing the denudation zone [143, 261]. The size of the denudation zone increases with energy input [40] and is dependent on the ambient gas pressure [143]. A significant denudation zone has to be avoided, as it can be a source of defects.

2.2.6 Defects in SLM

Different types of defects can occur in SLM parts such as pores, cracks, inclusions, segregations, residual stresses, or metallurgical imperfections [55, 255]. From the mentioned defects, porosity is the predominant one observed in the aluminum alloys studied in this thesis. Therefore, the three most important types of porosity (see Figure 2-21) and their formation are described below [1, 148, 256].

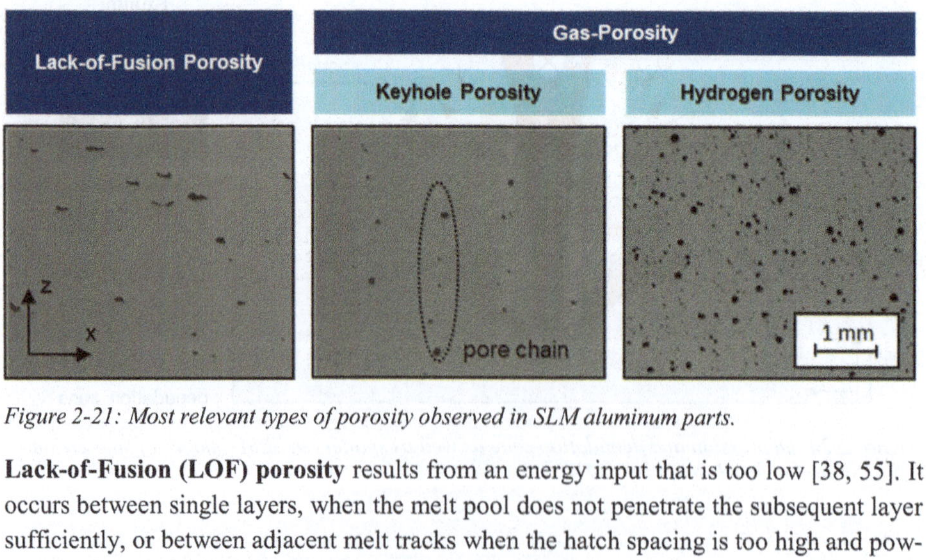

Figure 2-21: Most relevant types of porosity observed in SLM aluminum parts.

Lack-of-Fusion (LOF) porosity results from an energy input that is too low [38, 55]. It occurs between single layers, when the melt pool does not penetrate the subsequent layer sufficiently, or between adjacent melt tracks when the hatch spacing is too high and powder particles are not completely molten [143, 209, 255].

Gas-Porosity is a result of entrapped gases in the melt pool that are either caused by the evaporation of low-melting constitutes, dissolution of oversaturated elements (e.g., hydrogen), by pre-existent pores in the powder particles, or by the collapse of the melt pool (keyhole porosity) [17, 147, 235]. Two types of gas porosity are of particular importance for this thesis and described in more detail [1, 235]:

- **Keyhole porosity** occurs at high energy inputs when the welding mode changes from heat conduction to the undesired keyhole mode [20, 35]. In the keyhole mode, gas bubbles can emerge at the bottom of the melt pool when the keyhole collapses, as shown in Figure 2-22 [20, 221]. When the solidification time is not sufficient for the gas bubble to rise to the top, a spherical pore is formed in the solidified material [20]. Several studies show that keyhole porosity often (but not necessarily) occurs close to the laser turning points where the energy density is increased due to the acceleration or deceleration of the laser beam [20, 141, 255]. This can be identified by pore-chains, as shown in Figure 2-21. However, sky-writing can be used to avoid keyhole porosity at laser-turning points [141, 146].
- **Hydrogen porosity** is formed due to over-saturated hydrogen that dissolves in the melt pool during solidification or heat treatment. It particularly occurs in aluminum due to its high reactivity to hydrogen [1, 55, 235]. The hydrogen can either come from moisture on the surface of the powder particles or entrapped hydrogen in the particles [55]. As the hydrogen solubility is about 50 times higher for liquid compared to solid aluminum, and increases with temperature, hydrogen gas pores are formed during solidification in the rear section of the melt pool as shown in Figure 2-23 [85, 147, 235]. Depending on the direction of melt flow and the solidification speed, the gas pores can either rise to the top of the melt pool and escape or form spherical pores in the solid material [235]. However, hydrogen can also end up as super-saturated atoms in the crystal structure when the time for the dissolution is not sufficient. In this case, the hydrogen can form gas pores in the material when it is heated during the melting of neighboring tracks or heat treatment at sufficient temperatures [235].

It must be mentioned that it can be challenging to distinguish hydrogen and keyhole porosity by microscopic pictures, as they occur in similar size and shape. However, it can be assumed that keyhole porosity is present when patterns such as pore chains exist (see Figure 2-21) and that small spherical pores with a size $< 20\ \mu m$ are most probably hydrogen pores [55, 91, 138].

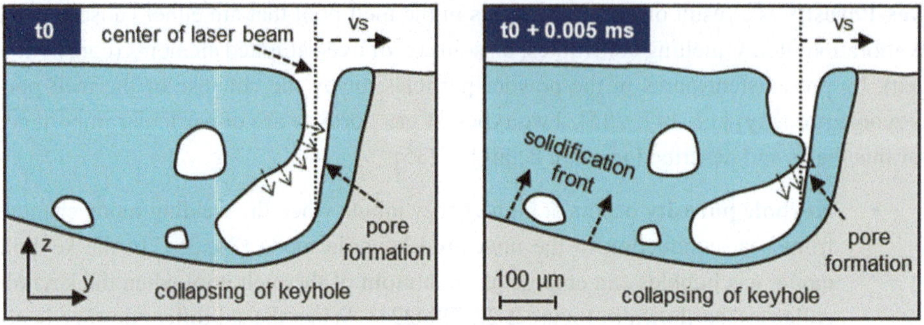

Figure 2-22: Pore formation due to the collapsing of a keyhole (based on [20]).

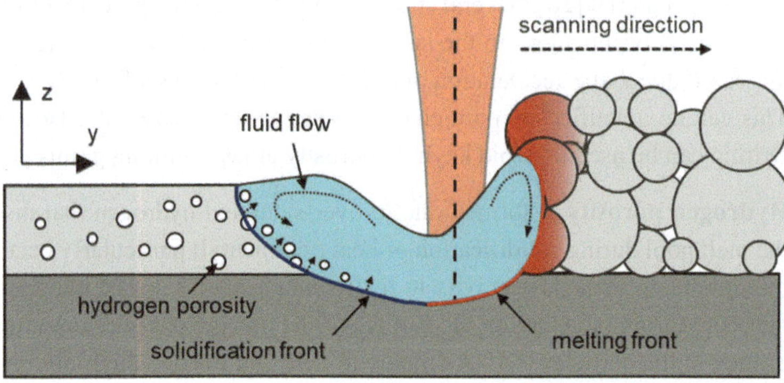

Figure 2-23: Formation of hydrogen porosity in the rear section of the melt pool (based on [235]).

2.3 Simulation of SLM and Related Processes

Significant effort has been made in recent years to simulate the SLM process. The numerical models are used to better understand the details of the SLM process and to accurately predict specific part characteristics, such as microstructure or residual stresses. The following gives an overview of the simulation of the SLM process (Chapter 2.3.1), lists relevant numerical models for SLM and related processes (Chapter 2.3.2), and summarizes the results (Chapter 2.3.3).

2.3.1 Overview of Simulation in SLM

Numerical models of SLM can be divided into three categories [48, 59, 148], as shown in Figure 2-24:

- **Micro-scale models** simulate the process on the scale of single powder particles (a few microns up to millimeters) [48]. Due to complexity and computational effort, it is not reasonable to simulate whole parts with these models. Instead, they concentrate on simulating single melt tracks or layers, increasing the physical understanding of the process and melt pool dynamics [48, 59]. Typical applications are the investigation of the laser-material interaction and optimization of process parameters [48, 139, 155]. Due to the speed and small scale in which the laser-material interaction occurs, these physics are difficult to investigate experimentally [20].

- **Macro-scale models** simulate the SLM process in the scale of millimeters to meters [59]. Usually, the models concentrate on simulating the thermomechanics of the process [139]. Details of the laser-material interaction and melt pool dynamics are ignored or incorporated in a simplified form. The overall level of detail and physical phenomena taken into account are significantly lower compared to micro-scale models [59]. A typical application of this kind of model is the prediction of residual stresses and distortion of SLM parts [139]. They can also be used to optimize the scanning strategy to minimize residual stresses [164, 170].

- **Multi-scale models** combine micro-scale and macro-scale models to ensure accurate predictions on all length scales while keeping the computational efforts reasonable [48]. Therefore, an excerpt of the domain is simulated with a high level of detail and the results are incorporated in the macro-scale model.

The classification of a numerical model to the categories is not always distinct and depends on the type of model, targeted application and modeling methods used. Table 2-1 gives an overview of different modeling methods and their typical application. For the numerical assessment of the melt pool dynamics and geometries, the Finite Volume Method (FVM) and Lattice Boltzmann (LB) method are most commonly used [203].

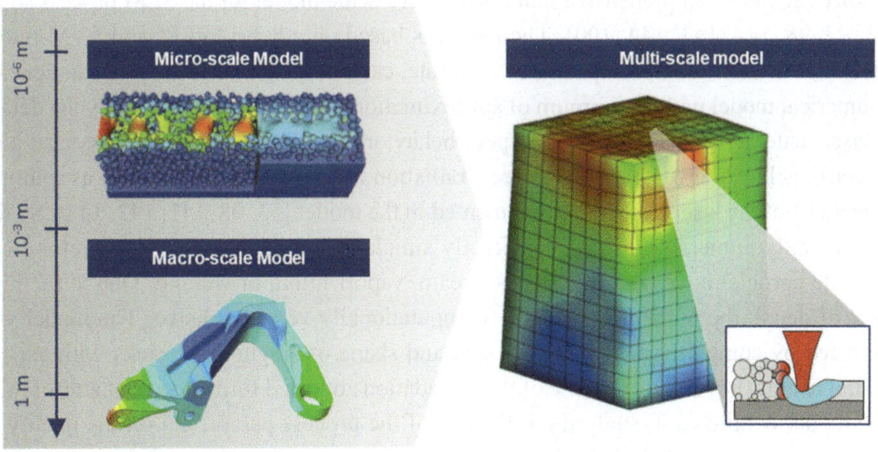

Figure 2-24: Different types of numerical models for SLM (graphics are from [163, 171]).

Table 2-1: Modeling methods used for the SLM process and their application (based on [139]).

Type of model	Modeling Method	Application	References
Micro-scale	Finite Element	Thermodynamics, mechanics	[61, 134, 174]
	Finite Volume / Volume of Fluid	Hydro- and thermodynamics	[36, 80, 160]
	Discrete Element	Powder bed generation	[175, 269]
	Monte Carlo	Beam absorption and ray tracing	[116, 266]
	Lattice Boltzmann	Hydro- and thermodynamics	[59, 105, 106]
	Phase Field	Hydrodynamics, microstructural evolution	[125]
	Cellular Automaton	Microstructural evolution	[257]
	Combination of above	Combination of above	[97, 120, 176]
Macro-scale	Finite Element	Thermodynamics, mechanics	[5, 164, 173]
Multi-scale	Combination of above	Combination of above	[48, 139, 148]

2.3.2 Relevant Numerical Models

Many numerical models have been developed and improved in recent years. This chapter gives an overview of the most relevant micro-scale models for SLM and related processes such as Laser Metal Deposition (LMD) and Laser Welding.

Khairallah, Matthews et al. from Lawrence Livermore National Laboratory, USA, developed the most comprehensive numerical micro-scale model for the SLM process so far [53, 97, 98, 141, 143, 144, 190]. The model is based on a hybrid FEM and FVM multiphysics code that was developed at the institute, called ALE3D. The goal was to develop a numerical model with a minimum of approximations that can be used to study the details of laser-material interaction and melt pool behavior. Therefore, all relevant physical phenomena such as thermo-capillary forces, radiation and vaporization, as well as multiple reflections of the laser beam are implemented in the model [53, 98, 141, 143, 144]. So far, it is the only numerical model that directly simulates the gas phase and, therefore, can simulate particle movement due to gas stream, vaporization, or wetting. Due to the high level of detail, the numerical model is computationally very expensive. The model was validated by comparing the melt pool size and shape of single melt tracks with experimental results [98] and by the size of the denudation zone and formation of spatters [143]. The model is utilized to study the influence of the process parameters on the quality of single melt tracks for SS316L. The results show that the formation of the melt pool is driven by surface tension and that the stochastic nature of the powder bed has a significant effect on the SLM process [97]. Additionally, the threshold energy that leads to the buildup of a keyhole during SLM is studied, which helps to identify the optimal processing window for SS316L [98]. Moreover, the effect of porosity at the end of scan tracks due to the collapse of keyhole is investigated and mitigation strategies are proposed [141]. The results demonstrate that the depletion of metal powder particles in the surrounding of a melt track and formation of spatters during SLM is a result of the "competition between outward metal vapor flux directed away from the laser spot and entrainment of powder particles in a shear flow of gas driven by a metal vapor jet [143]." Finally, the model is used to better understand the absorption of the laser energy for different materials and process parameters [190] and to numerically investigate the effect of process parameters and other laser beam profiles on the microstructural evolution [53, 144].

Dai, Gu et al. from Nanjing University of Aeronautics and Astronautics, China, developed a multi-scale model for SLM using the commercial software ANSYS [24, 37, 73]. The micro-scale model concentrates on simulating the melt pool dynamics and densification behavior, while the macro-scale model simulates the temperature- and stress-development of the multi-layer SLM process [36, 73]. The micro-scale model utilizes the FVM and the interface of the fluid is described by the Volume of Fluid (VOF) method. The powder is represented as a continuum material with modified material properties in earlier versions of the model [36, 37], but is modeled explicitly in more recent models [73, 244]. The material properties are temperature-dependent and the capillary forces, as well as recoil pressure due to vaporization, are taken into account. At the same time, multiple reflections of the laser beam are neglected. The fluid is treated as an incompressible Newtonian fluid and the material flow is assumed to be laminar. The model is validated by comparing the porosity of experimental and numerical results for different LEDs [244]. With the help of the model, the densification behavior of a copper matrix composite is investigated [37], the effect of process parameters on the surface quality of aluminum composite is simulated [36] and thermodynamics in the melt pool during the processing of nickel-based superalloys is studied [244]. Among other things, the results show that surface tension plays a vital role in the formation of the melt pool [36] and that the Marangoni flow increases the heat and mass transfer significantly [244] and can cause the entrapping of gas bubbles in the melt pool [37].

Leitz, Singer et al. from Plansee SE, Austria, developed a numerical model using the commercial software COMSOL Multiphysics [123–126]. The model is based on FEM and uses the Phase-Field method to define the interfaces [125]. The powder layer consists of spherical and monomodal particles and temperature-dependent material properties are implemented as well as surface tension and capillary forces [125, 126]. The fluid is treated as an incompressible Newtonian fluid and the flow of material is considered to be laminar [125, 126]. An angle-dependent absorption of the laser beam is implemented. At the same time, shadowing effects and multiple-reflections are neglected [125]. Furthermore, the effect of vaporization and radiation are modeled explicitly [125]. The model is validated by comparing the melt pool width for different processing parameters with experimental results [125]. The numerical model is used to better understand the effect of process parameters on the SLM process for molybdenum and steel [124–126].

Markl, Körner et al. from the University of Erlangen-Nürnberg, Germany, developed a numerical model to simulate the SLM and EBM process [9, 15, 105, 106, 198]. The code has recently been transferred into a stand-alone software called SAMPLE [140]. The numerical model is based on the Lattice-Boltzmann method [15] and the surface is described by the Volume of Fluid method [106]. Simulations are performed in 2D [15, 105] and 3D [9]. Surface tension, capillary forces, vaporization and multiple-reflections are considered in the model [140, 183]. A raindrop model is used to generate a random powder layer [105, 140]. The model is validated by comparing the cross-section of single melt tracks for different process parameters to experimental results [105] and the quality of molten layers qualitatively [9]. The simulation results show that the packing density in the powder layer as well as the randomness of the powder bed significantly influence the melt pool characteristics and quality of the built [105]. Moreover, it is shown that low scanning speeds are favorable for a stable melt pool [106]. Additionally, the effect of process parameters on melt pool lifetime and geometry is studied to determine the process window for a stable melting process [198].

Wei, Chen et al. from the State Key Laboratory for Manufacturing Systems Engineering, China, developed a numerical model based on FVM to simulate the SLM process for AlSi10Mg, nickel-base superalloy K418 as well as TiAl6V4 [176, 234, 245]. The model is implemented in the commercial software Flow3D. The fluid and gas phase interface is described by the Volume of Fluid method [234]. The fluid is implemented as an incompressible Newtonian fluid and the flow is considered to be laminar [176]. Surface tension, capillary forces, effects of vaporization and temperature-dependent material properties are included in the model. The powder bed is generated by simulating the application of powder particles using the Discrete Element Method (DEM) module of Flow3D [176]. The model is validated by comparing the width and depth of single melt racks with experimental results [234] and the morphology of single tracks qualitatively [245]. The model is primarily used to investigate the effect of process parameters on the melt pool dynamics. One of the key findings is that the Marangoni flow is significantly dependent on process parameters and that high scanning speeds tend to cause melt instabilities and balling [176, 234, 245].

Lee, Zhang et al. from Ohio State University, USA, developed a Finite Volume model similar to *Wei, Chen et al.* described above, using the commercial software Flow3D [120, 121]. The model is used to quantitatively study the effect of process parameters on the geometry of single melt tracks and the balling effect for Inconel 718. One of the key findings is that the balling tendency can be reduced using smaller powder particles and lower scanning speeds [120].

Cheng, Tuffile et al. from Robert Bosch LLC, US, and Robert Bosch GmbH, Germany, developed a numerical model to study the effect of the powder packing density on the melt pool using the commercial software Flow3D [26–28]. The software's DEM module is used to simulate different powder application methods and their effect on the powder bed characteristics [26]. The resulting powder beds are used as an input for the simulation of single tracks for the tool-steel H13, SS316L and TiAl6V4 [27, 28]. The results show that a lower powder bed packing density results in deeper and more narrow melt pools, that the heat transfer in the melt pool is governed by convective rather than conductive heat transfer [26] and that unstable keyholes are the main cause for the formation of porosity [27, 28].

Loh, Song et al. from Nanyang Technological University, Singapore, developed a numerical model to simulate the melt pool dimension for the aluminum alloy AA6061 [134, 135, 214]. The model is based on FEM and implemented in the commercial software COMSOL Multiphysics. The powder is represented by a continuous material with specific properties to increase the numerical efficiency of the models [134, 135]. The flow of material and, therefore, the effect of capillary forces and surface tension are modeled indirectly. The volume shrinkage due to the melting of the powder is implemented by the element birth and death technique and energy losses due to vaporization are considered by a negative heat source. The model is validated by comparing the melt pool width and depth for different processing parameters with experimental results [134, 135]. With the model's help, the effect of the process parameters on the melt pool dimensions and vaporization are simulated [134]. The results show, among other things, that preheating the powder by a second laser beam can be an effective method to reduce vaporization in the SLM process [135].

Qiu, Panwisawas et al. from the University of Birmingham, UK, have developed a numerical model to study the effect of the process parameters on the surface structure and porosity for SLM of TiAl6V4 and nickel-based super-alloys [19, 172, 181]. The model is based on FVM and implemented in the open-source CFD software OpenFOAM. The interface between the metal and gas is calculated by the VOF method. Important phenomena, such as heat losses due to vaporization, vapor pressure and surface tension are included in the model, while multiple reflections of the laser beam are neglected [181]. Temperature dependent material properties are implemented and a random powder bed is generated using a raindrop model [19, 172]. It is assumed that the molten material is incompressible and that the fluid flow is laminar [181]. Due to a fine mesh, the model can simulate the formation of spatters [181]. The numerical results show that unstable melt flow is a major cause for porosity and that Marangoni flow, as well as recoil pressure due to vaporization, are the primary cause for this [181]. Additionally, it is found that a finer particle size distribution reduces the amount of micro-voids [19].

Megahed, Mindt et al. from the ESI Software Company, Germany, have developed a multi-scale numerical model for the SLM process in the fluid dynamics software suite CFD-ACE+ [48, 148, 157, 268]. In this model, the temperature distribution and residual stresses from the micro-scale serve as an input for the macro-scale predicting residual stresses and deformations of complete parts [48, 157]. The micro-scale model is based on FEM and utilizes the VOF method to describe the fluid-gas interface [48]. The model incorporates important physical phenomena such as (thermo-)capillary forces and vaporization. Randomly distributed powder particles are implemented and the fluid is assumed to be incompressible and to behave like a Newtonian fluid [157, 268]. The micro-scale model is validated by comparing the size of the single melt tracks for different processing parameters to experimental results [48]. The macro-scale model, on the other hand, is validated by comparing the calculated distortion of specimens to experimental results [48, 148]. With the numerical model, the researchers analyzed the effect of different scanning strategies on the residual stresses and distortion of coupons [48] and the effect of powder bed characteristics on the SLM process [268].

Fan, Zhang et al. from Missouri University of Science and Technology, USA, have developed a 3D numerical micro-scale model to simulate the Laser Metal Deposition (LMD) process for TiAl6V4 and pre-mixed powder materials [53, 129, 260]. With the numerical model, the influence of process parameters and the powder flow rate on the melt pool dynamics are investigated [53, 129]. The model is based on FVM and utilizes the VOF method to calculate the liquid-gas interface. Temperature-dependent material properties are implemented as well as important physical phenomena such as (thermo-)capillary forces and vaporization. The liquid is considered to be an incompressible Newtonian fluid [53, 260]. A mass-source term is applied to the surface to simulate the addition of powder during the process. An attenuation function is implemented to account for the effects of the laser powder interaction and the powder temperature rise during the deposition of the particles [53]. The model is validated by comparing melt pool size, peak temperatures and occurrence of lack-of-fusion defects for different process parameters to experimental results [53]. Among other things, the results show that the argon gas flow rate has a significant effect on the separation of pre-mixed metal powders [129].

Lin, Wang et al. from Shanghai Jiao University, China, have developed a detailed numerical model to simulate the laser-material interaction and melt flow in laser remote welding of aluminum alloys and zinc-coated steel [39, 83, 131]. The model is based on FVM, utilizes the VOF method to calculate the free surface flow and is implemented in the commercial software Flow3D [83, 131]. The model considers the (thermo-)capillary forces and the fluid is assumed to be incompressible and behave like a Newtonian fluid [131]. The material properties are temperature-dependent. The effect of vaporization pressure on the melt pool and multiple-reflections of the laser beam inside the keyhole are considered [83, 131]. The model is validated by comparing the occurrence and shape of keyhole induced porosity [131] as well as the size and shape of the weld beads [83] to experimental results for different processing parameters. The simulation results reveal that the violent melt flow behind the keyhole is a major reason for gas bubble formation and that the ambient gas pressure has a significant effect on the collapsing of the keyhole [83, 131].

Saldi, Kleijn et al. from the Technical University of Delft, Netherlands, have developed a three-dimensional numerical model to simulate the Marangoni-driven free surface flow during laser spot welding [100, 194, 195]. The model utilizes the FVM, simulates the free surface flow by the VOF method and is implemented in the open-source CFD software OpenFOAM. The model considers the effect of (thermo-)capillary forces and vaporization [194, 195]. The fluid is modeled as an incompressible Newtonian fluid and attention is paid to accurately describe the phase transition between solid and liquid and its effects on the volume shrinkage. The numerical model is validated by comparing the free surface and the melt flow to experimental results from the literature [194] and the weld pool size to experimental results [195]. With the model, the effect of the process parameters and surface-active elements on the Marangoni flow and shape of the melt track after solidification is studied [194, 195]. Among other things, the results demonstrate that the Marangoni flow has a strong influence on the fluid flow and formation of instabilities in the weld pool during laser spot welding [194].

2.3.3 Summary

Table 2-2 provides an overview of the described numerical models and their main characteristics. It can be seen that the important physical phenomena that govern the heat and mass transfer in the melt pool, such as thermo-capillary forces and vaporization effects, are considered in most numerical models. The only numerical model that directly models the gas phase has been developed by Khairallah, Matthews et al. from the Lawrence Livermore National Laboratory. This model enables the simulation of powder particle movement due to gas stream, vaporization or wetting at the costs of numerical efficiency. However, although the explicit simulation of the gas phase allows for additional insights, it is not necessary for an accurate simulation of the essential melt pool dynamics, as demonstrated by the other researchers.

Furthermore, it can be seen that the Finite Volume Method, in combination with the Volume of Fluid method, is the most commonly used technique for the micro-scale simulation of SLM and related processes in the mentioned numerical models. The reasons for this are:

- FVM is especially suitable for Computational Fluid Dynamics (CFD) and more accurate in modeling fluid flow than FEM [162],
- VOF has advantages over other free surface modeling techniques like the Level-Set or the Phase Field method in terms of mass-conservation [41, 194],
- FVM and VOF have few constrictions on mesh-geometries [162],
- The topology of the surface can change without any constrictions as VOF works with a fixed mesh. This is a significant advantage for the simulation of particle-melting or spatters formation [41].

Based on the outcomes, it can be concluded that FVM and VOF are very suitable for modeling the effect of process parameters on the melt pool dynamics in SLM. Furthermore, numerical models such as the one of Wei, Chen et al. [176, 234, 245], Lee, Zhang et al. [120, 121] or Cheng, Tuffile et at. [26–28] provide a good framework for the envisaged simulation of this thesis as they have implemented all relevant physical phenomena. However, these models still lack proper validation for the simulation of the SLM process with aluminum alloys and have not been used to simulate different laser beam profiles. This, in combination with the fact that the models are not available publicly or commercially, makes it necessary to set-up a new numerical model.

Table 2-2: Overview of numerical micro-scale models for SLM and related processes.

Authors and References	Simulation Software	Modelling Method (on Micro-Scale)	Physical Phenomena						Different Laser Beam Profiles	Validated for Aluminum Materials
			Temp. Dependent Material Prop.	Effects of Vaporization	Single Particles	(Thermo-)capillary Forces	Multiple Reflections	Particle Entrainment and Spatter		
Khairallah, Matthews et al. [53, 97, 98, 141, 143, 144]	ALE3D	FEM/FVM	+	+	+	+	+	+	o	-
Dai, Gu et al. [24, 36, 37, 73, 244]	ANSYS / FLUENT	FVM/VOF	+	+	+	+	-	-	-	+
Leitz, Singer et al. [123–126]	Comsol Multiphys.	FEM/PF[1]	+	o	+	+	-	-	-	-
Markl, Körner et al. [15, 105, 106, 139, 140, 183]	SAMPLE	LB[2]/VOF	+	+	+	+	+	-	-	-
Wei, Cheng et al. [176, 234, 245]	Flow3D	FVM/VOF	+	+	+	+	+	-	-	-
Lee, Zhang et al. [120, 121]	Flow3D	FVM/VOF	+	+	+	+	+	-	-	-
Cheng, Tuffile et al. [26–28]	Flow3D	FVM/VOF	+	+	+	+	+	-	-	-
Loh, Song et al. [134, 135, 214]	Comsol Multiphys.	FEM	+	o	-	o	-	-	-	+
Qui, Panwisawas et al. [19, 172, 181]	OpenFOAM	FVM/VOF	+	+	+	+	-	-	-	-
Megahed, Mindt et al. [48, 148, 157, 268]	CFD-ACE+	FEM/VOF	+	+	+	+	+	-	-	-
Fan, Zhang[3] [53, 129, 260]	not specified	FVM/VOF	+	+	x	+	x	-	-	-
Lin, Wang et al.[3] [39, 83, 131]	Flow3D	FVM/VOF	+	+	x	+	+	-	-	+
Zaki Saptari Saldi[3] [100, 194, 195]	OpenFOAM	FVM/VOF	+	+	x	+	x	-	-	-

+	fully implemented	o	partially implemented
-	not implemented	x	not applicable / no information provided
1	Phase Field method	2	Lattice Boltzmann method
3	numerical model used to simulate LMD or laser welding		

2.4 Laser Beam Shaping

Laser beam shaping describes the process of altering the spatial or irradiance profile of the laser beam into the desired shape at a specific location behind the shaping element [43, 230]. Different beam shaping techniques are classified and described in Chapter 2.4.1. Chapter 2.4.2 gives an overview of laser beam shaping in SLM and related processes.

2.4.1 Laser Beam Shaping Techniques

Laser beam shapers can be divided into three categories [43, 230]:

- **Apertures** are used to filter a specific portion of the beam while the rest is absorbed or reflected [230]. Only the beam's contour is changed while the intensity of the part that passes the aperture remains unchanged. As a result, a significant amount of energy is lost. This is the reason why some do not consider apertures as real beam shapers [43]. Due to the significant energy loss and limited beam-shaping capabilities, apertures are not suitable for beam shaping in high power laser-material processes [230].

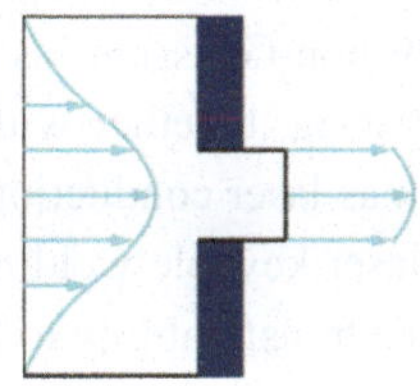

- **Field mappers** transform an input beam to an output beam with a desired spatial and irradiance profile [43, 230]. The advantage of this beam-shaping technique is the high optical efficiency as field mappers are nearly lossless. Additionally, a small focus diameter can be obtained (< 1.5 times of the diffraction-limited size [117]). Field mappers need a well-defined single-mode laser beam as input and the desired beam profile is obtained at a specific distance from the focusing optic [43, 117, 118].

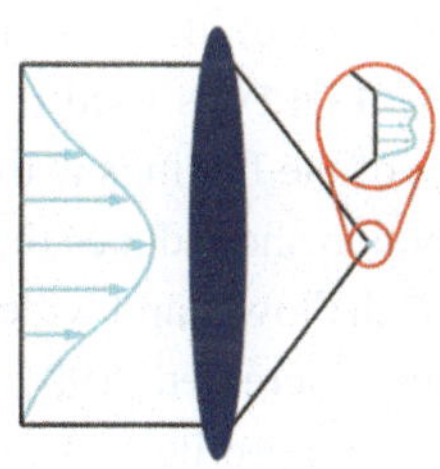

- **Homogenizers** (also: beam integrators) break up the input laser beam into multiple single beams which are then focused by a primary optic [43, 230]. The desired beam shape is obtained by an overlap of the multiple focused laser beams. Homogenizers create the desired profile at a specific distance from the focusing element. In contrast to field mappers, homogenizers do not require a well-defined single-mode laser beam as input and the size of the focused beam is significantly larger (usually > 4 times of the diffraction-limited focus size) [43, 118].

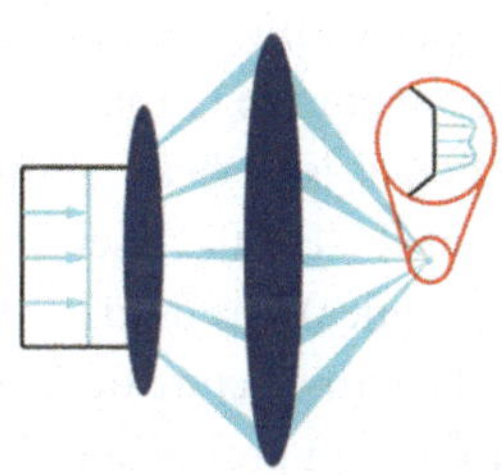

The beam shaping elements used in field mappers and homogenizers can consist of refractive or diffractive optics [43, 230]. Both types can have a high optical efficiency and focusability [84]. On the one hand, a diffractive optical element is less sensitive to alignment errors of the input laser beam can create virtually any laser beam profile and is usually more compact in size [111]. On the other hand, a refractive optical element is not dependent on the wavelength of the laser, but has some limitations regarding the possible output laser beam profile [111, 117]. As a small focus diameter and high optical efficiency are desired for SLM, field mappers can be considered the best option for beam shaping in this application.

2.4.2 Laser Beam Shaping in SLM and Related Processes

Only minor research has been done so far to investigate how the SLM process can benefit from non-Gaussian laser beam profiles. The most relevant studies are presented in the following, together with studies on the laser beam profile's effect on related processes such as laser conduction welding and Laser Metal Deposition (LMD). Studies that focus on laser keyhole welding and the processing of non-metallic materials are not presented as the transferability to SLM is limited.

Roehling, Matthews et al. from Lawrence Livermore National Laboratory, USA, numerically and experimentally investigated the effect of alternative laser beam profiles on the shape of single tracks as well as micro-structure in SLM [144, 186, 190, 208]. For the numerical investigations, the model described in Chapter 2.3.2, developed by Khairallah, Matthews et al., is utilized [207]. For the experiments a diffractive field mapper from HOLO/OR was used that alters the Gaussian input beam to a donut beam profile [190]. One of the findings is that a donut beam profile leads to a more uniform temperature distribution and reduces the keyhole effect as well as the amount of spatters [190]. Additionally, shallower and wider melt pools are achieved and the energy efficiency of the process can be increased [190]. Furthermore, it is found that the grain structure is dependent on the beam profile and can be changed by adopting the ellipticity of the laser beam [144, 186, 208]. Based on this, grain-structure engineering is suggested for future research by adopting the laser beam profile "on the fly" [186, 208].

Zhirnov, Gusarov et al. from Moscow State University of Technology, Russia, investigated the influence of the laser beam profile on single tracks and microstructure for cobalt-chromium [153, 166–168, 179, 265]. The single-track experiments were done for three different laser beam profiles (Gaussian, top-hat and donut), each with a different focus diameter. For the laser beam's modulation, a refractive field mapper (focal-piShaper from AdlOptica) was used [153, 168]. Among others, the researches demonstrate that the shape and microstructure of the single tracks obtained at different power densities depend on the laser beam profile and that non-Gaussian laser beam profiles can reduce the denudation zone [166, 168]. Additionally, a better metallurgical bonding and more uniform melt tracks are obtained with a top-hat laser beam compared to a Gaussian one [167, 168].

Furthermore, it is shown by numerical investigations that a donut beam profile can significantly reduce the vaporization for SS316L [179].

Loh, Yeong et al. from Nanyang Technological University, Singapore, investigated the effect of the laser beam intensity profile on the width and depth of single melt tracks for the aluminum alloy AA6061 [133]. Two different laser beam intensity profiles were used: a Gaussian profile with a focus diameter of 80 μm and a top-hat profile with a focus diameter of 730 μm [133]. The results show that the melt width is significantly larger for the top-hat laser beam profile, while the penetration depth is similar to the one of the Gaussian beam. Based on this, it is suggested to use both laser beam profiles simultaneously to increase the productivity in the SLM process (see hull-core strategy, explained in Chapter 2.2.4) [133].

Cloots et al. from Inspire AG, Switzerland, explored the influence of the laser beam intensity profile on the microstructure, porosity and crack formation in SLM for the nickel-based superalloy IN738LC [33, 34]. The investigations were done for two different laser beam profiles, a Gaussian profile with a focus diameter of 100 μm and a donut profile with a focus diameter of 140 μm. The donut beam profile was obtained by a diffractive field mapper (vortex lens from HOLO/OR) [34]. The results show that the average porosity is higher and the crack density is lower for the donut profile. The primary reason for the higher porosity is seen in the bonding defect, which could be related to lower energy densities of the donut profile [33, 34]. Furthermore, no keyhole formation is observed for the donut profile for any parameter tested [34]. Additionally, the influence of the laser beam profiles on the crystalline orientation and grain boundary structures was found to be minor [33, 34].

Hansen, Kristiansen et al. from IPU Technology Development, Denmark, experimentally investigated the possibility to control the weld pool width and depth in laser welding of steels by laser beam shaping [84, 94, 108]. The laser beam intensity profile was altered by a diffractive optical beam integrator to obtain different patterns of Gaussian profiles. The results show that line patterns obtain significantly wider melt pools and that the melt pool width and depth can be controlled for laser keyhole as well as laser conduction welding [84]. It is demonstrated that the welding efficiency is significantly higher for line patterns compared to similar weld beads obtained with a defocused Gaussian beam [84]. The authors suggest using line patterns to improve the surface quality and robustness of the laser welding process and to control the mixing in laser welding or laser metal deposition of dissimilar materials [84]. Additionally, the potential to weld several beads in parallel by beam shaping is validated on a butt-joint configuration [94, 108].

Sundqvist from Luleå University of Technology, Sweden, investigated the influence of the laser beam profile on the weld seam shape for laser spot welding in heat conduction mode by numerical simulation and experiments [219]. The investigations were done for laser spot welding of thin stainless-steel plates with three different laser beam profiles: a Gaussian, top-hat, and C-shaped beam profiles. For the experiments, a customized diffractive optical field mapper was used to obtain the desired intensity profiles [219]. The results show that the weld shape can directly be influenced by the laser beam profile, while a direct correlation between beam shape and weld geometry is not possible due to the complex melt flow. Therefore, an analytical model is developed to predict the weld shape for spot welding of thin metal plates for different laser beam profiles [219].

Han, Liou et al. from University of Missouri, USA, numerically investigated the laser beam profile's influence on the melt pool in heat conduction spot welding [82]. With the help of a 3D FEM model, the melt pool is simulated for four different profiles: Gaussian, donut and, cylindrical and rectangular TEM_{01} (Transverse Electromagnetic Mode) profiles. The results show that the laser beam profile significantly influences the melt pool shape and flow pattern. The Gaussian beam profile has the highest aspect ratio, making it ideal for deep penetration welding, cutting or drilling. The non-Gaussian profiles show a significantly wider and shallower melt pool, which offers advantages in low-intensity laser material processes such as Laser Metal Deposition [82].

Kell, Tyrer et al. from Loughborough University, UK, studied the possibility to control the weld pool shape in laser conduction welding and heat absorption in wire-fed laser cladding by customized beam profiles obtained by diffractive optical field mappers [67, 68, 86, 95]. It is demonstrated that the shape of the weld bead and microstructure are directly influenced by a square or peaked edge beam in laser conduction welding [86, 95]. Furthermore, that beam profiles with a more even energy distribution result in better wetting behavior and less dilution in wire-fed cladding [67, 68].

Tseng et al. from National Chung Cheng University, Taiwan, numerically investigated the effect of tailored laser beam profiles on the heat distribution and clad bead profile in LMD [226]. The model was implemented in the SYSWELD FEM software. The simulations were done for a cobalt clad on a steel base plate with a Gaussian and different multimode laser beam profiles [226]. The numerical results show that the laser beam profile directly influences the temperature distribution, melt pool size and clad beam geometry. For non-Gaussian laser beams, a shallower melt pool and a smaller heat-affected zone are obtained [226].

2.4.3 Summary

Table 2-3 summarizes the results described in Chapter 2.4.2. The results demonstrate that the laser beam profile can directly influence the melt pool shape and microstructure in SLM and related material processes such as laser conduction welding and laser cladding [34, 208, 230]. For SLM in particular, only minor research has been done and for most of the experimental work, the beam size was not constant within the experiments [34, 190, 252, 265]. This limits the meaningfulness of the results as the observed differences can come from both the beam size and the beam shape. Additionally, little attention has been paid to the effect of the beam profile on stability and productivity in SLM [153].

In addition to the above, it has to be noted that the author of this thesis has published work on the potential of beam shaping for AlSi10Mg that demonstrates that a donut beam profile can lead to a larger process window, fewer defects, and higher feasible energy input for AlSi10Mg, compared to a Gaussian beam profile [239]. However, a detailed analysis of the causes (e.g., by numerical results), as well as investigations on the effect of the beam profile on productivity, have not been published yet.

Table 2-3: Summary of effects of the laser beam profile on SLM and related processes.

Authors	Process	Major Findings
Roehling, Matthews et al. [144, 153, 186, 190, 208]	SLM	- process window, shape of single tracks and microstructure are dependent on the beam profile - possibility to engineer microstructure by adopting the laser beam profile - vaporization, spatters and denudation are reduced by a donut beam profile, compared to a Gaussian beam profile - smoother, more shallow and wider melt tracks are obtained by a donut beam profile, compared to a Gaussian beam profile
Zhirnov, Gusarov et at. [153, 166–168, 179, 265]	SLM	- shape of single tracks and microstructure depend on the laser beam profile - improved metallurgical bonding and reduced denudation zone with a top-hat beam profile, compared to a Gaussian beam profile - reduction in vaporization by a donut beam profile, compared to a Gaussian beam profile
Loh, Yeong et al. [252]	SLM	- increased melt pool width and similar penetration depth for a top hat beam profile, compared to a Gaussian beam profile
Cloots et al. [33, 34]	SLM	- higher porosity and lower crack density for a donut profile, compared to a Gaussian beam profile - avoidance of keyhole effect by a donut beam profile - limited influence of beam profile on crystalline orientation and grain boundary structure for IN738LC
Hansen, Kristiansen et al. [84, 94, 108]	Laser Welding	- control of weld pool geometry in laser welding of steels by application of laser beam line patterns - increase in welding efficiency by line pattern
Sundqvist. [219]	Laser Welding	- the beam profile can directly influence weld bead profile in laser conduction welding
Han, Liou et al. [82]	Laser Welding	- direct influence of beam profile on the weld pool shape and flow pattern - more shallow and wider melt pools for non-Gaussian laser beam profiles
Kell, Tyrer et al. [67, 68, 86, 95]	Laser Welding & LMD	- control of the shape of the fusion zone is possible by laser beam shaping in laser conduction welding - better wetting and less dilution in LMD with a non-Gaussian laser beam profile
Tseng et al. [226]	LMD	- shallower melt pool for non-Gaussian beam profiles - heat-affected zone is smaller for a non-Gaussian beam profile - the beam profile can directly influence clad beam geometry

3 Numerical Model

Due to the small scale and high dynamic of the SLM process, it is complicated to investigate some important phenomena experimentally, such as Marangoni flow, melt velocities and the magnitude of vaporization. Therefore, a numerical micro-scale model is developed to gain more in-depth insight into the laser beam profile's effect on the SLM process, based on the state of the art presented in Chapter 2.3.

The following sections describe the general considerations taken into account for the development of the model (Chapter 3.1). Furthermore, the set-up of the model (Chapter 3.2), the method to analyze the results (Chapter 3.3) and the validation of the developed model (Chapter 3.4) are presented. A detailed description of the numerical model is provided in Appendix A.1.

3.1 General Considerations

To analyze the relevant melt pool characteristics numerically, the most important physical phenomena have to be implemented accurately. As described in Chapter 2.2.5, the dynamics of the melt pool are governed by the (thermo-)capillary forces as well as vaporization [98, 101, 189]. At the same time, other phenomena have a minor influence, such as buoyancy force and radiation [98, 130]. For the development of the model, the following assumptions and simplifications have been made:

- The fluid behaves like an incompressible, Newtonian fluid. This assumption is accurate for most of the domain with limitations at the surface between the gas and fluid phases due to the significant difference in density [87]. Additionally, this assumption becomes inaccurate for velocities higher than half of the sound of speed, which is not expected in the melt pool [169]. The effect on the overall melt pool dynamics due to this simplification can be considered as negligible [114].

- The melt pool flow is turbulent as the Reynolds numbers are expected to be $Re >$ 1,000 (see Equation 3.1). Since vaporization pressure can be seen as a significant disturbance to the natural fluid flow, turbulence occurs for Reynolds numbers of $Re > 100$ [194]. This assumption of a turbulent fluid flow is a difference to most other numerical models that assume a laminar fluid flow.

- The gas phase is not directly modeled and, therefore, displacement of single particles e.g., by vaporization or wetting forces, are ignored.

- The effect of the vapor induced shear stress on the melt pool dynamics in heat conduction welding is low compared to other surface forces and, therefore, is ignored.

T. M. Wischeropp, *Advancement of Selective Laser Melting by Laser Beam Shaping*,
Light Engineering für die Praxis, https://doi.org/10.1007/978-3-662-64585-7_3

- The simulated material is isotropic and homogeneous. Although the simulated material is an alloy and, hence, this assumption is not true on the scale of nanometers, the effect of the anisotropy and inhomogeneity of the material can be seen as insignificant considering the scale of the simulation (scale of micrometers).

$$Re = \frac{\rho_{Al} \cdot v_m \cdot w_l}{\mu_{Al}} > \frac{2{,}500\ kg/m^3 \cdot 2\ m/s \cdot 0.0002\ m}{0.001\ kg/(m \cdot s)} > 1{,}000 \tag{3.1}$$

ρ_{al}: Density of aluminum-alloy; estimated with ~2,500 kg/m³ [254]
v_m: Maximum velocity of fluid; estimated with > 2 m/s [97, 181]
l_{c_mp}: Characteristic length; estimated with ~200 µm (width of melt pool) [234]
μ: Dynamic viscosity of fluid; estimated with ~0.001 kg/(m·s) [234]

Flow-3D (V11.2.6.4) from Flow Science Inc. is selected as the software for the simulation of the SLM process. It has successfully been used to simulate the melt pool dynamics by Wei, Chen et al. [176, 234, 245], Lee, Zhang et al. [120, 121] and Cheng, Tuffile et at. [26–28] (see Table 2-2 on page 36). However, the researchers did not investigate the processing of aluminum or the processing with a non-Gaussian beam profile. The software is based on the Finite-Volume-Method (FVM) and utilizes the Volume-of-Fluid (VOF) method to calculate the liquid-gas interface. The advantages of FVM with VOF are its mass-conservativeness, that it can be applied to any geometrical mesh and the fact that there are no restrictions on topological changes of the fluid surface [41]. These are all important factors for the envisaged simulation.

A detailed description of the numerical model and the implementation of the important phenomena, such as the heat and mass transfer at the surface, laser absorption, vaporization and solidification can be found in Appendix A.1.

3.2 Setup of Numerical Model

The following describes the set-up of the domain, its boundary conditions, the powder layer and the material properties implemented in the model.

3.2.1 Domain and Boundary Conditions

The model concentrates on simulating single tracks with a length of 2.5 mm to keep the computational efforts reasonable. The length is defined as all three characteristic sections of a single melt track occur for this length, as shown in Figure 3-1: "Start," "steady-state" and "end" sections. To have a considerable distance between the region of interest (melt pool) and the boundary of the simulated domain, the size of the solid material is set to 3 mm · 1mm · 0.4 mm. At high energy inputs, the thickness of the solid material is increased up to 0.7 mm to keep a minimum distance of 250 µm between the bottom of the melt pool and the domain's boundary.

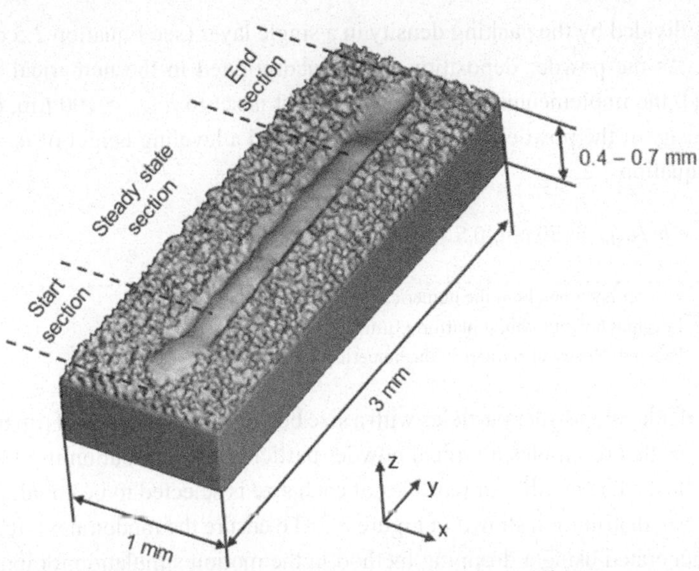

Figure 3-1: Sections of single melt tracks and dimensions of the simulated domain.

A continuative boundary condition is defined at the boundaries 1 to 5 (see Figure 3-2). This implements zero normal derivatives for all quantities and guarantees a smooth heat transfer at the boundary [57]. A specific pressure condition is implemented at boundary 6 to keep the ambient gas pressure constant during the simulation. A pressure of 10,150 Pa is used as this resembles the pressure in the build chamber of the SLM machine used for the experiments. Furthermore, a mesh size of 15 μm is defined based on the results of a mesh convergence study presented in Appendix A.2.

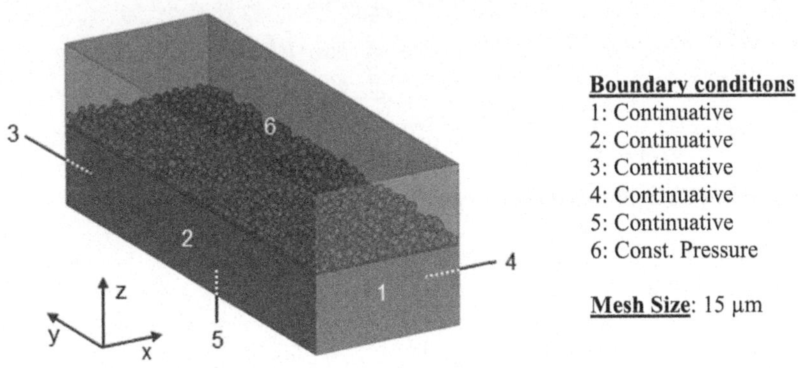

Figure 3-2: Overview of boundary conditions of the simulated domain.

3.2.2 Implementation of Powder Bed

To accurately describe the melt pool dynamics of the SLM process, it is essential to implement a realistic powder layer. When the depletion of powder particles due to vaporization is neglected, the powder layer height will equal the leveling height of the building platform h_l, divided by the packing density in a single layer (see Equation 2.5 on page 18) [149, 164]. As the powder deposition is not incorporated in the numerical model (see Chapter 4.1), the implemented powder layer height is set to h_{l_sim} = 100 µm, assuming a packing density of the powder layer of a_{sim} = 50% and a leveling height of h_l = 50 µm, as shown in Equation 3.2.

$$\mathrm{h}_{l\_sim} = h_l / a_{sim} = 50\ \mu m / 0.5 = 100\ \mu m \tag{3.2}$$

h_{l_sim}: Powder layer height in the numerical model [µm]
h_l: Leveling height of build platform [µm]
a_{sim}: Packing density of powder in the numerical model [1]

For the powder layer, powder particles with a size between 25 – 65 µm (10 µm increments) are defined, as this resembles a typical powder particle size distribution used in SLM for aluminum alloys. The number of particles of each size is selected to be equal, resulting in the particle size distribution shown in Figure 3-3. To ensure the randomness of the powder layer, it is generated using a dropping method in the motion simulation toolbox of *SolidWorks®* [180]. The resulting powder layer is shown in Figure 3-4 and has a packing density of 49.8%.

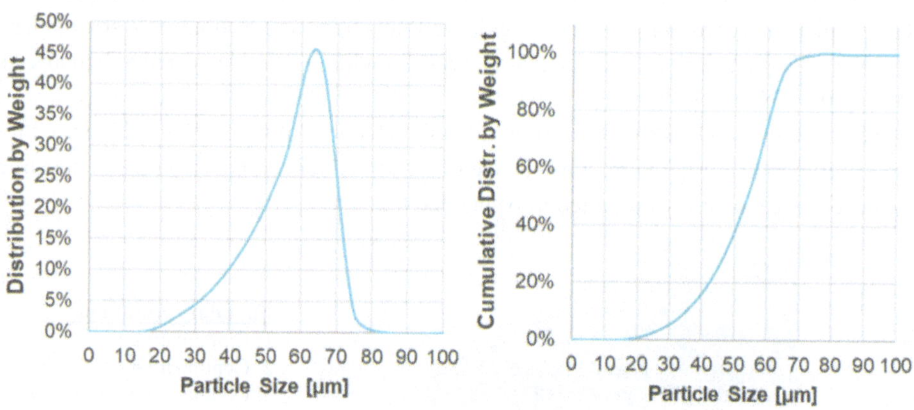

Figure 3-3: Particle size distribution of the powder layer in the numerical model.

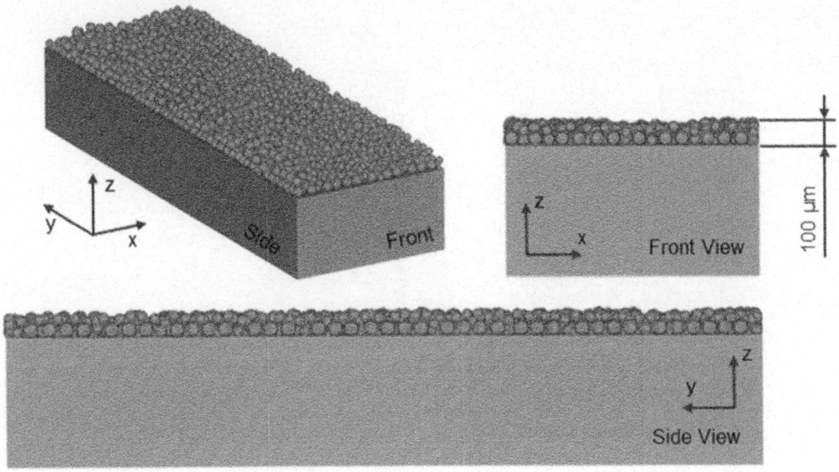

Figure 3-4: Overview of powder layer implemented in the numerical model.

3.2.3 Material Properties

The numerical investigations are primarily done for the aluminum alloy AlSi10Mg, which is the most commonly used aluminum alloy in SLM [225]. The temperature-dependent density ρ_{Al} as well as the viscosity μ_{Al} are shown in Figure 3-5 and the heat conductivity k_{Al} and heat capacity C_{p_Al} in Figure 3-6. All other relevant material properties are listed in the following Table 3-1.

Table 3-1: Material properties of AlSi10Mg implemented in the numerical model.

Property	Value	Source
Density ρ_{Al}	see Figure 3-5	[176]
Viscosity μ_{Al}	see Figure 3-5	[176]
Heat capacity C_{p_Al}	see Figure 3-6	[176]
Thermal conductivity k_{Al}	see Figure 3-6	[176]
Solidus temperature T_{ST_Al}	830.15 K	[145]
Liquidus temperature T_{LT_Al}	869.15 K	[145]
Boiling temperature T_{b_Al}	2,767 K	[75]
Boltzmann constant σ_B	$5.67 \cdot 10^{-8}$ W/(m²K)	[176]
Latent heat of fusion ΔH_{F_Al}	389 J/g	[145]
Latent heat of vaporization ΔH_{V_Al}	10,900 J/g	[79]
Absorptivity of liquid a_{Al_l}	40%	[176]
Absorptivity of solid a_{Al_s}	24%	[21]
Young's modulus E_{Al}	$71 \cdot 10^9$ N/m²	[145]
Surface tension σ_{Al}	$1.00726 - 0.000152 K^{-1} \cdot T$ N/m	[176]
Wetting angle θ_{0_Al}	155 °	[47]
Convective heat transfer coefficient h_{ce_Al}	5,000 W/(m²K)	[51]

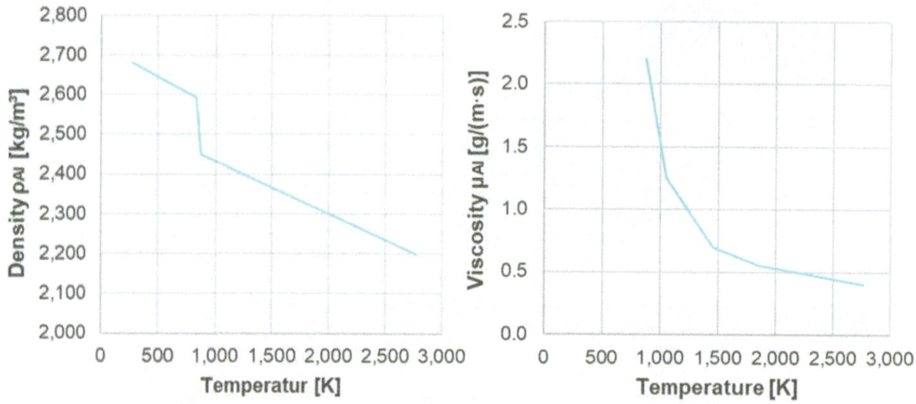

Figure 3-5: Temperature-dependent density ρ_{AL} and viscosity μ_{Al} for AlSi10Mg [176].

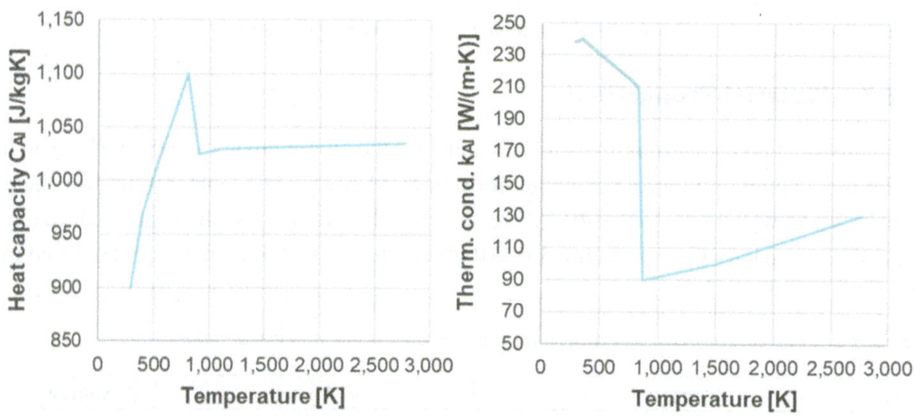

Figure 3-6: Temperature-dependent heat capacity C_{p_Al} and thermal conductivity k_{Al} for AlSi10Mg [176].

3.3 Evaluation of Numerical Results

Flow3D enables the visualization of the numerical results but has limitations in the quantitative analysis. Therefore, the numerical results are exported and analyzed by a MATLAB script that is written to determine the relevant melt pool characteristics quantitatively. The following gives an overview of the relevant attributes and how they are analyzed. To account for statistical variations within the process, each parameter is averaged over five different positions along the scan track (y_1 to y_5), as shown in Figure 3-7. If more or fewer positions are analyzed, it is explicitly stated.

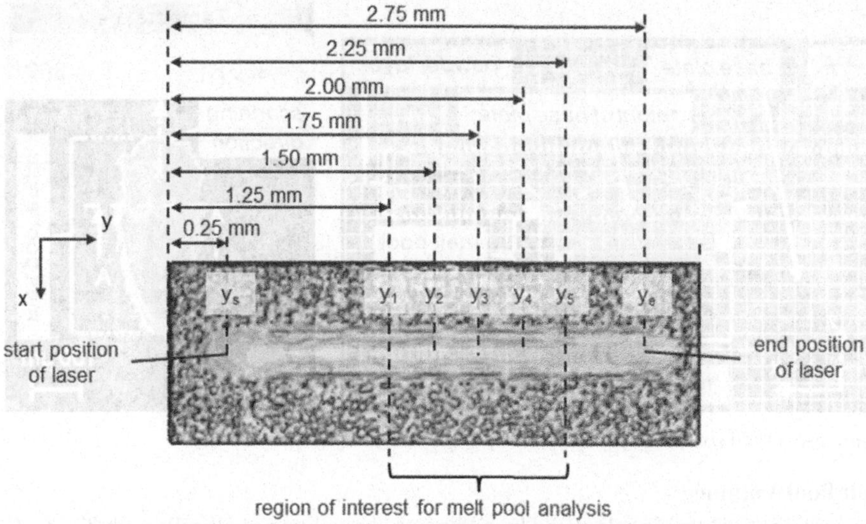

Figure 3-7: Scan track with laser-positions and regions of interest (y_s: starting position of the laser beam; y_1: start of the region of interest; y_5: end of the region of interest; y_e: end-position of the laser beam).

Melt Pool Width, Depth and Length

The melt pool width and depth are important parameters that, among other things, are used to validate the numerical model (see Chapter 3.4). Furthermore, the melt pool length is a stability-criterion, as melt pools that are too long lead to balling [98, 110].

For the melt pool width and depth analysis, only the part of the melt pool in the solid material (excluding the powder layer) is considered, as shown in Figure 3-8. The melt pool width is calculated by subtracting the minimum x-coordinate of a cell with temperature T above solidus temperature T_S from the maximum x-coordinate of a cell with $T > T_S$. The melt pool depth is calculated by subtracting the cell's z-position with the minimum z-coordinate with $T > T_S$ from the height of the solid material. The melt pool length is calculated by subtracting the minimum y-coordinate of a cell with temperature T above solidus temperature T_S from the maximum y-coordinate of a cell with $T > T_S$. If a cell's temperature is between solidus and liquidus temperature ($T_S < T < T_L$), the contribution of the specific cell to the width, depth, or length is calculated based on a linear interpolation between T_S and T_L.

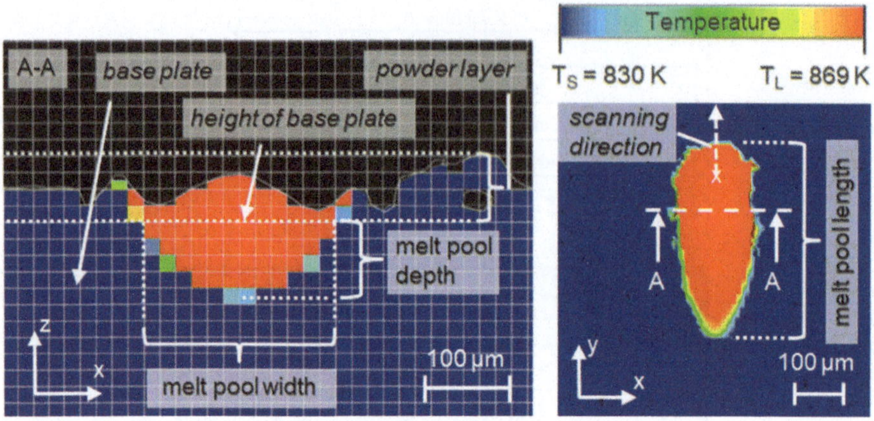

Figure 3-8: Analysis of melt pool width, depth and length of single tracks.

Melt Pool Volume

The melt pool volume is calculated by adding up the volume of all cells with $T > T_S$. Cells that have a fluid fraction F between $0 < F < 1$ are considered corresponding to their fluid fraction coefficient (e.g., cells with a fluid fraction $F = 0.5$ are considered with half their volume). Furthermore, cells with a temperature between solidus and liquidus temperature $T_S < T < T_L$ are considered based on a linear interpolation between T_S and T_L.

Thermo-Capillary Flow / Surface Tension Gradient

The magnitude of the thermo-capillary flow (Marangoni flow) depends on the surface tension gradients (see Chapter 2.2.5). To quantitatively describe the strength of the Marangoni flow, the sum of the surface tension gradients is calculated. Therefore, the surface tension gradient of each melt pool surface cell is calculated by averaging the difference between its own surface tension to that of its neighboring cells.

Melt Pool Velocities

The average magnitude of the fluid-velocity is calculated along with its standard deviation to describe the melt velocities quantitatively. Furthermore, the flow-directions are analyzed by evaluating the direction-dependent components of the flow vectors of each cell.

Vaporization Pressure / Vapor Pressure Force

Vaporization induces pressure on the surface of the melt pool. The total vapor pressure force is calculated to get quantitative metrics for the magnitude of the vaporization in the process. Therefore, each surface cell's vapor pressure is multiplied by its surface area and the value is summed up for all melt pool surface cells.

3.4 Validation of Numerical Model

Single-track experiments are done and the width and depth of the tracks are compared to those of the simulation to validate the numerical model, as exemplarily shown in Figure 3-9. This method is commonly used to validate numerical models in SLM [134, 203, 234] and it follows the standards for measuring weld geometries provided in DIN EN ISO 5817 for conventional welding [44]. The investigations are done for two laser beam profiles (Gaussian and donut) with the parameters listed in Table 3-2. More details about the set-up of the single-track experiments are laid out in Chapter 5.1.

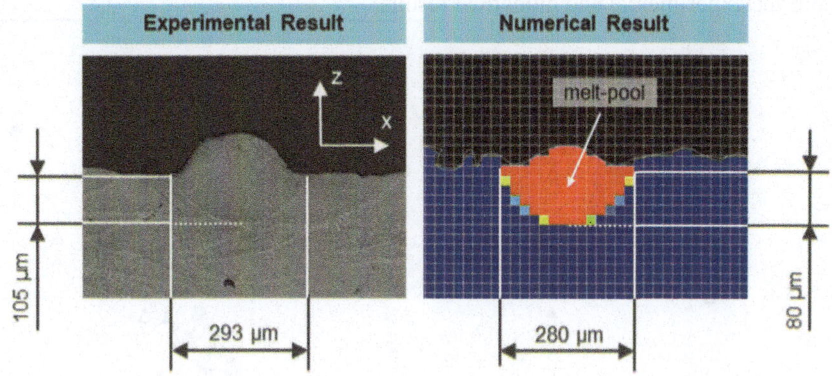

Figure 3-9: Measured width and depth of one cross-section of experimental and numerical single track, manufactured with the donut beam profile, a laser power of P_L = 500 W and scanning speed of v_s = 1,000 mm/s.

Table 3-2: Parameters used for the validation of the numerical model.

Factor	Level
Laser power P_L [W]	400; 500; 600; 700
Scanning speed v_S [mm/s]	1,000; 1,500; 2,000; 2,500
Leveling height of build platform h_l	50 µm
Temperature of build platform T_{BP}	20 °C
Laser beam profile	Gaussian (G), donut (D)

The single track's width and depth from the experiments and simulation are shown in Figure 3-10 for the Gaussian and in Figure 3-11 for the donut beam profile.

For the Gaussian beam profile, the width increases in a logarithmic manner with an increase in linear energy density (LED), while the depth increases linearly. This is in agreement with data from other researchers [78, 234]. The numerical model shows the same trends, while the width seems to be slightly overestimated for low LEDs and underestimated for high LEDs. However, the difference is in the range of the standard deviation of the experimental results. The melt pool depth, on the other hand, is overestimated for all LEDs and the gradient is slightly steeper in the numerical model. The difference in the melt pool depth can partly come from the negligence of powder depletion in the numerical model, and the difference in powder layer height between the numerical model and experiments. As discussed in Chapter 2.2.4, the actual powder layer height will be higher than the 100 µm implemented in the simulation due to powder depletion in the experiments [156, 238]. This can contribute to a reduced melt pool depth in the experiments.

For the donut beam profile, a similar behavior is seen. The melt pool width predicted by the numerical model is in very good agreement with the experimental results, while the depth is slightly higher. Comparing the results to the ones obtained for the Gaussian beam profile, one can see that the width is similar for both beam profiles. At the same time, the linear progression in melt pool depth is about 20% higher for the Gaussian beam profile for both the experimental and numerical results.

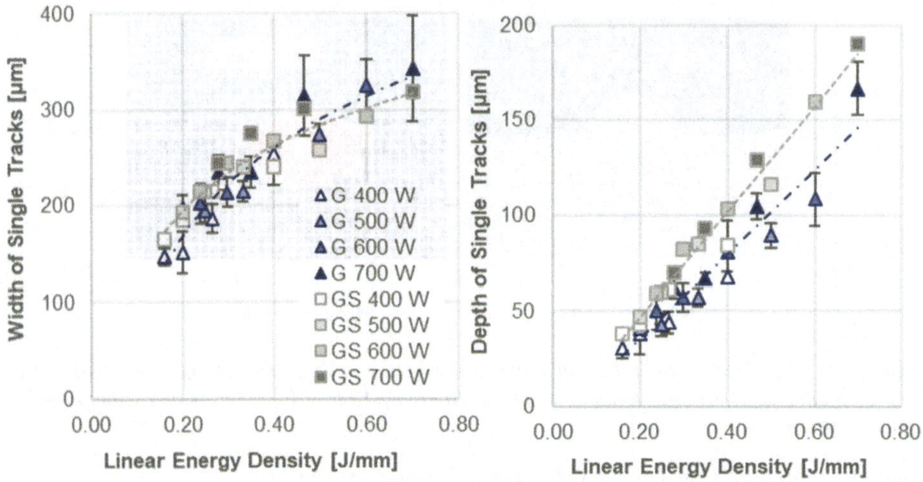

Figure 3-10: Comparison of width and depth of experiments (G) and simulation (GS) for the Gaussian beam profile.

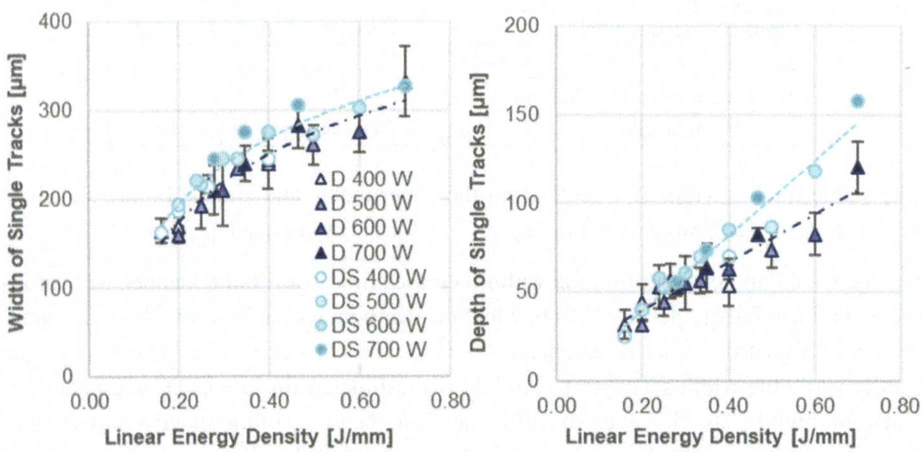

Figure 3-11: Comparison of width and depth of experiments (D) and simulation (DS) for the donut beam profile.

Overall, it can be stated that the numerical model agrees with experimental results when it comes to predicting the melt pool width but overestimates the melt pool depth to a certain degree. However, for the analysis in this thesis, the relative difference in the results obtained between the Gaussian and the donut beam profile is of particular importance. As this is very similar between the experimental and numerical results (~20% higher melt pool depth for the Gaussian beam profile), the numerical model can be used to compare the beam profiles' effect on the governing melt pool dynamics.

4 Experimental Set-Up

The following sections describe the design of experiments (Chapter 4.1) as well as experimental equipment (Chapter 4.2). Furthermore, the powder material characteristics (Chapter 4.3) and method for spatter analysis (Chapter 4.4) are laid out.

4.1 Design of Experiments

In this thesis, the effect of two different laser beam profiles, a standard Gaussian and a donut beam (see Figure 4-3), on the SLM process is investigated. For this, single-track, as well as density cube experiments are performed:

- **Single-track experiments** are performed to understand the effect of the laser beam profile on the size and shape of single melt tracks in the SLM process.
- **Density cube experiments** are performed to analyze the effect of the laser beam profile on the multi-layer SLM process. The experiments are carried out to understand the effect of the beam profile on the process stability (feasible energy input, process window and spatter formation) and productivity (build rate).

The single-track experiments are carried out with AlSi10Mg and the density cube experiments with AlSi10Mg as well as AlSi16Sc. Two different aluminum alloys are used to investigate whether the beam profile's effect depends on the alloy composition.

Single-Track Experiments

In the experiments, single tracks are produced with different combinations of laser power and scanning speed and the appearance, width, and depth are analyzed by micro-sections. Table 3-2 (see page 51) shows the full-factorial design of the experiments. Five single tracks are analyzed for each combination of process parameters to take into account statistical variations of the process.

As written in Chapter 2.2.4, the powder layer height in SLM increases with the first layers until it reaches a *steady state*. The *steady state* value of the powder layer height depends on the leveling height of the build platform, the packing density of powder, and process parameters. To perform the single-track experiments and to compare them to each other, the same powder layer height for all single tracks must be ensured. To do so, the single tracks are built on cubic specimens with a building height that guarantees that the steady-state condition for the powder layer height is reached (> 20 layers). When the last layer of the density cube is molten, the build platform is lowered and an additional layer of powder is applied for the single tracks. Before the single tracks are molten, the specimens are allowed to cool down for more than 30 minutes to reach room temperature. Figure 4-1 illustrates the set-up of the single-track experiments.

T. M. Wischeropp, *Advancement of Selective Laser Melting by Laser Beam Shaping*,
Light Engineering für die Praxis, https://doi.org/10.1007/978-3-662-64585-7_4

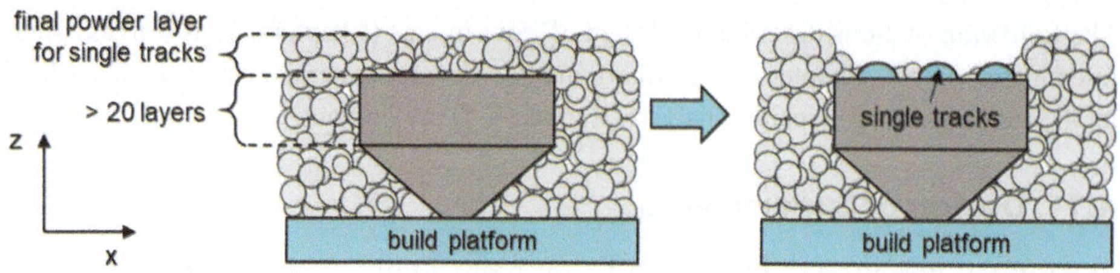

Figure 4-1: Schematic overview of the set-up of the single-track experiments (not to scale).

Density Cube Experiments

The density cube experiments are performed to understand the effect of the beam profile on the multi-layer SLM process. In a first test series, a full-factorial design of experiments is conducted with the parameters listed in Table 4-1 for AlSi10Mg and AlSi16Sc. A second test series with the parameters listed in Table 4-2 is conducted to increase the investigated process window size. It is set-up to include tests with the maximum scanning speed and laser power available on the test rig. The second test series is done for AlSi10Mg, only due to the non-availability of a sufficient amount of AlSi16Sc powder. Additionally, high-speed videos with 1,000 frames per second are recorded to analyze the spatter formation for the two different beam profiles for the experiments done with AlSi10Mg. For reference, Table 4-3 lists the standard process parameters used on the test rig for both materials.

Due to the high number of specimens (254), a total of twelve build jobs are manufactured. To verify the consistency between the build jobs, several specimens are manufactured twice on different build jobs. Micro-sections are prepared to analyze the porosity and defects within the specimens. To account for statistical variations within the density cubes, three micro-sections are prepared for each density cube.

Table 4-1: Parameters of full-factorial experimental design of the first series of density cube experiments.

Factor	Level
Laser power P_L [W]	300; 400; 500; 600; 700; 800
Scanning speed v_S [mm/s]	1,000; 1,500; 2,000
Hatch distance h_S [mm]	0.1; 0.15; 0.2
Leveling height of build platform h_l	50 µm
Temp. build platform T_{BP}	200 °C
Laser beam profile	Gaussian (G), donut (D)
Laser beam diameter	140 µm

Table 4-2: Parameters of full-factorial experimental design of the second series of density cube experiments.

Factor	Level
Laser power P_L [W]	600; 700; 800; 900
Scanning speed v_S [mm/s]	2,000; 2,500; 3,000
Hatch distance h_s [mm]	0.2; 0.25
Leveling height of build platform h_l	50 µm
Temp. build platform T_{BP}	200 °C
Laser beam profile	Gaussian (G), donut (D)
Laser beam diameter	140 µm

Table 4-3: Standard parameters used on the test rig for the processing of AlSi10Mg and AlSi16Sc.

Factor	AlSi10Mg	AlSi16Sc
Laser power P_L	370 W	300 W
Scanning speed v_S	1,300 mm/s	1,600 mm/s
Hatch distance h_s	0.19 mm	0.1 mm
Leveling height of build platform h_l	60 µm	50 µm
Temp. build platform T_{BP}	200 °C	200 °C
Laser beam profile	Gaussian (G)	Gaussian (G)
Laser beam diameter	85 µm	85 µm
Build rate b ($v_S \cdot h_s \cdot h_l$)	12.5 mm³/s	8 mm³/s

4.2 Equipment

Below, a description of the equipment used for the experiments is provided.

SLM Machine: AconityLAB from Aconity GmbH

The experiments are conducted on an AconityLAB from Aconity3D GmbH, Germany, with a Coherent FL1000R 1 kW fiber laser from Coherent LaserSystems GmbH, Germany, and a customized optical bench. This enables the integration of a beam shaper in the optical path of the laser beam as shown in Figure 4-2. The technical specifications of the machine are listed in Table 4-4. It has to be mentioned that the maximum available laser power was 900 W due to power losses as a result of the aging of the laser source.

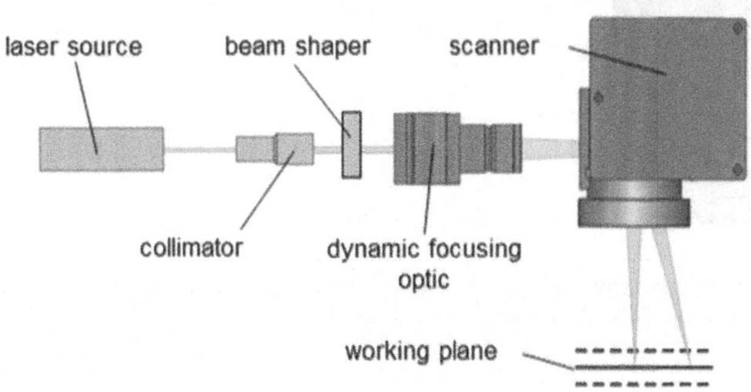

Figure 4-2: Set-up of the customized optical bench (based on [197]).

Table 4-4: Technical specification of AconityLAB machine with customized optical bench.

Diameter of the build platform	Ø 170 mm
Max. build height	200 mm
Focus size d_f	80 – 500 µm
Temp. build platform T_{BP}	Max. 500 °C
Max. scanning speed v_S	3 m/s
Max. laser power P_L	900 W

Beam Shaper: M-Shaper from Holo/Or

The Coherent FL1000R 1 kW fiber laser installed in the AconityLAB has a Gaussian (TEM_{00}) beam profile with a beam quality of $M^2 < 1.05$ [187]. The focused beam size is approximately 85 µm. To change the beam profile from Gaussian to donut, an M-Shaper from Holo/Or LTD, Israel, is installed in the optical path, as shown in Figure 4-2 [90]. The M-Shaper is a diffractive field mapper that creates a donut beam profile in the focus point and increases the $1/e^2$ diameter by approximately 65% (in this case from d_f ~85 µm to d_f~140 µm). For the Gaussian beam profile, the laser is defocused to obtain the same diameter in the working plane. This is a significant difference to other beam-shaping investigations done in SLM by Zhirnov et al. [264], Metel et al. [153] or Cloots et al. [34], for example, where the size of the beam was not constant between the different beam profiles. This limits the validity of the results since the differences can come from both the beam profile or the beam size. The measured beam profiles are shown in Figure 4-3.

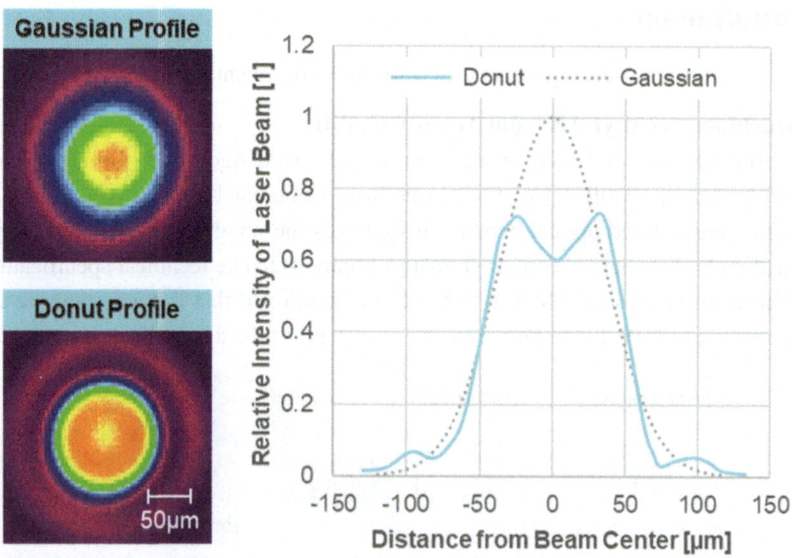

Figure 4-3: Measured beam profiles of the donut and defocused Gaussian beam (color-scale not consistent between images on the left).

Equipment for Laser Beam Analysis

For the measurement and calibration of the laser power, a 1000W-BB-34 thermal sensor, for the measurement of the beam shape, a SP928 camera, and for the measurement of the focus position along the z-axis, a BeamWatch AM is used. All measurement devices are from Ophir Optronics LTD, Israel. The measurements of the beam power, profile and position are done before each experiment to guarantee the correct calibration of the laser and optics.

Equipment for Material and Specimen Analysis

For the analysis of appearance and chemical composition of the powder a Scanning Electron Microscope (SEM) (JSM 5910LV) from Joel GmbH, Germany, and an energy-dispersive X-ray spectroscopy machine (EDX) (INCAX-ACT 51-ADD006) from Oxford Instruments plc, UK, is used. The particle size distribution is measured with the Camsizer X2 from Retsch Technology GmbH, Germany.

The micro-sections for the single tracks and porosity analysis are prepared with a Struers Tegramin-30 grinding machine. The porosity is analyzed with a digital microscope VHX-5000 from Keyence Corporation, Japan.

4.3 Material

Two different aluminum alloys are used in the experiments. AlSi10Mg is selected as the primary material, since it is the most commonly used aluminum alloy in SLM [240]. Additionally, density cube experiments are performed with a hyper-eutectic, high-strength AlSi16Sc aluminum alloy, to investigate if the beam profile's effect depends on the alloy composition. The AlSi16Sc material was developed within the European Horizon2020 project *BionicAircraft* (GA: 690689).

AlSi10Mg Powder

Figure 4-4 shows an SEM picture and the measured particle-size distribution of the AlSi10Mg powder used in the experiments. The powder particles are mostly spherical, with some satellites, and have a particle size distribution of 26-60 µm. The powder was supplied by SLM Solutions Group AG, Germany, with a specified particle size distribution of d_{10} = 20 µm and d_{90} = 63 µm.

Table 4-5 shows the powder's chemical composition, determined by EDX analysis, which is the result of the measurement of three samples. The measured amount of silicon in the material is slightly higher than specified (maximum of 11%), which can partly result from the normalization as other potential alloying elements have not been measured (e.g., Fe, Cu, ...). Similar results for the chemical composition measurement by EDX have been reported by other researchers [2]. However, the higher silicon amount can affect the mechanical properties and solidification behavior. Moreover, it reduces the hydrogen solubility, which can increase the porosity in the parts [2, 85, 147].

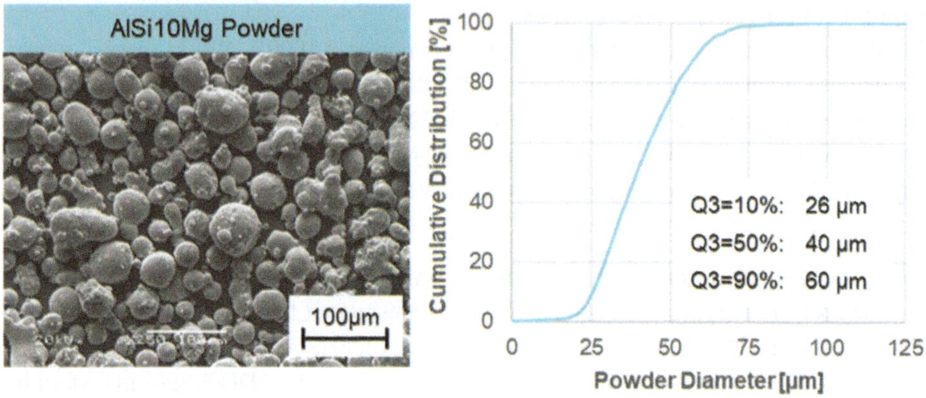

Figure 4-4: SEM picture and measured particle size distribution of the AlSi10Mg powder used in the experiments.

Table 4-5: Chemical composition of AlSi10Mg powder analyzed by EDX (normalized for main alloying elements).

	Al	Si	Mg	Total
AVG	86.61%	12.94%	0.45%	100%
STD	0.53%	0.54%	0.03%	-
Specification	balance	9 – 11%	0.2 – 0.45%	100%

AlSi16Sc Powder

Figure 4-5 shows an SEM picture and the measured particle size distribution of the AlSi16Sc powder used in the experiments. The powder was supplied by Tekna Plasma Europe, France, with a specified particle size distribution of d_{10} = 20 µm and d_{90} = 63 µm. The AlSi16Sc powder particles are mostly spherical and have a similar particle size distribution, similar to the AlSi10Mg powder. Overall, the distribution is slightly broader, with a higher number of particles with a diameter < 25 µm and > 60 µm. As the difference is relatively small, an effect on the processability due to different particle morphology, compared to the AlSi10Mg powder, is not expected.

Table 4-6 shows the powder's measured chemical composition, determined by EDX analysis, which is the result of the measurement of three samples. As for AlSi10Mg, the amount of silicon is slightly higher than defined in the specification, which can partly result from the normalization of the measured alloying elements.

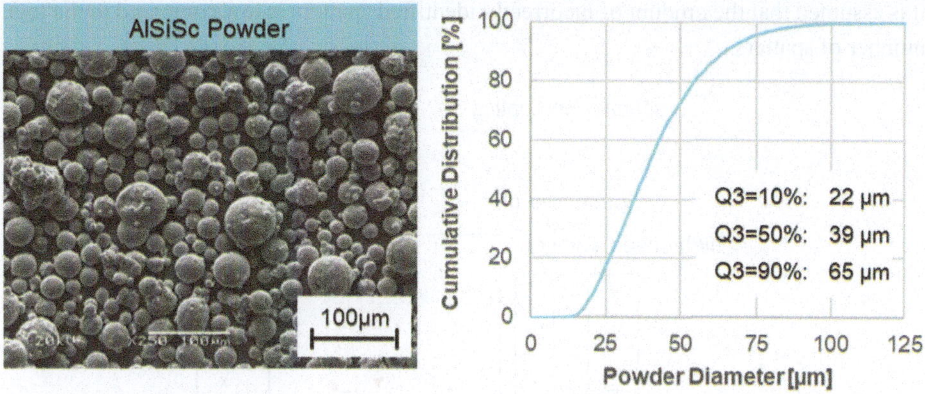

Figure 4-5: SEM picture and measured particle-size distribution of the AlSi16Sc powder used in the experiments.

Table 4-6: Chemical composition of AlSi16Sc powder analyzed by EDX (normalized for alloying elements).

	Al	Si	Sc	Total
AVG	81.34%	18.18%	0.48%	100%
STD	0.23%	0.29%	0.12%	-
Specification	balance	15 – 17%	0.3 – 0.5%	100%

4.4 Analysis of Spatters

As described in Chapter 2.2.5, spatters lead to instabilities in the process and can cause defects when they fall onto the powder bed. High-speed videos with 1,000 frames per second are recorded during the manufacturing of AlSi10Mg density cubes to investigate the beam profile's influence on the amount of spatters. The experimental set-up is shown in Figure 4-6. The amount of spatters is analyzed with the help of a MATLAB script that automatically converts every single image of the video into binary images, as shown in Figure 4-7. The binary images make it possible to count the amount of spatters within each frame automatically. For the analysis, average values are calculated over 200 images to account for variations in the process.

The picture is split up into an image with *bright* and an image with *dark* spatters for processing. *Bright* spatters are defined as spatters with a brightness above a defined threshold, and *dark* spatters below this threshold. This is done to analyze the ratio of spatters ejected from the melt pool (bright spatters) to spatters resulting from entrainment of powder particles (dark spatters). More details on the definition of this threshold and processing of the images are provided in Appendix A.3. However, it must be mentioned that the correlation between bright spatters and spatters that are ejected from the melt pool is not distinct. This is due to the circumstance that spatters ejected from the melt pool can become dark spatters when they cool down and due to the possibility that entrained powder particles can become bright spatters when they are molten due to an interference with the laser beam. However, it is assumed that the amount of incorrectly identified spatters is low compared to the total number of spatters.

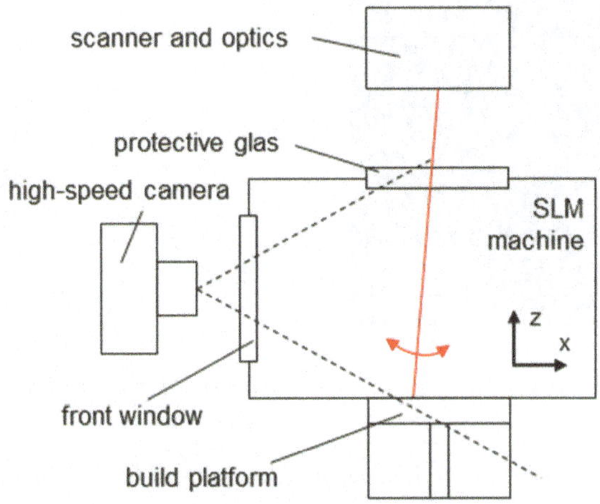

Figure 4-6: Experimental set-up for the recording of the spatters.

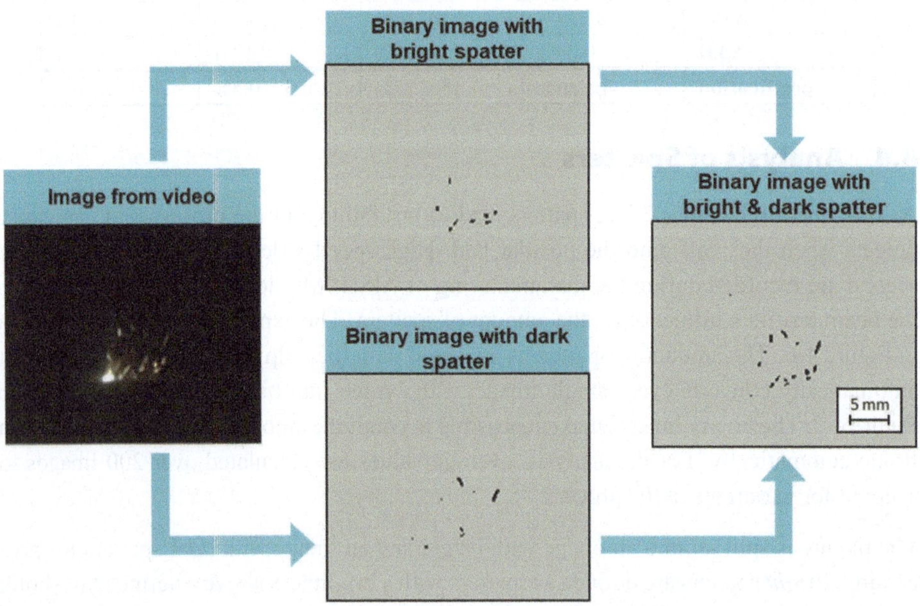

Figure 4-7: Image of a video with spatters and processed binary images with identified bright and dark spatters.

5 Effect of Laser Beam Profile on SLM Process

In this chapter, the effect of the laser beam profile on the SLM process is evaluated. In a first step, the effect on the melt pool and its key characteristics is numerically analyzed (Chapter 5.1). Afterward, the numerical results are used to discuss the experimental results obtained for single tracks (Chapter 5.2) and multi-layer SLM experiments (Chapter 5.3) for AlSi10Mg. Furthermore, the effect on the processing of a high-strength AlSi16Sc aluminum alloy is laid out (Chapter 5.4).

5.1 Numerical Assessment of Effect on Melt Pool

The melting process in SLM is complex, with several interrelated physical mechanisms taking place simultaneously, as described in Chapter 2.2.5. The phenomena governing the melt pool dynamics are the recoil pressure due to vaporization and the (thermo-)capillary forces [132, 143, 181, 203]. As the experimental evaluation of the governing phenomena is challenging, they are analyzed numerically. Therefore, the melting of single tracks is simulated with the numerical model described in Chapter 3 and the key characteristics are analyzed numerically. Afterward, the outcomes are used to explain the experimental results laid-out in Chapters 5.2 to 5.4.

5.1.1 Melt Pool Size and Shape

The correct melt pool size and shape are crucial for a stable SLM process. The melt pool penetration has to be high enough to achieve a bonding to the subsequent layer, the width has to be sufficient to merge with adjacent melt tracks and the length-to-width ratio should be small enough to avoid balling (see Chapter 2.2.5). Additionally, the melt pool should be stable in order to avoid defects [22, 181, 223].

Figure 5-1 exemplarily shows different melt pool shapes produced by the Gaussian and donut beam profiles for selected process parameters during the melting of single tracks. It demonstrates that the melt pools are overall more elongated and deeper for the Gaussian beam profile. The width, on the other hand, is higher for the Gaussian beam at a laser power of $P_L = 300$ W and lower for all other parameters shown. The reason for this can be found in the limited penetration of the melt pool produced by the donut beam profile at the laser power of $P_L = 300$ W, and the circumstance that only the part of the melt pool in the base plate is considered for the calculation of the melt pool width (see Figure 3-8 on page 50).

T. M. Wischeropp, *Advancement of Selective Laser Melting by Laser Beam Shaping*,
Light Engineering für die Praxis, https://doi.org/10.1007/978-3-662-64585-7_5

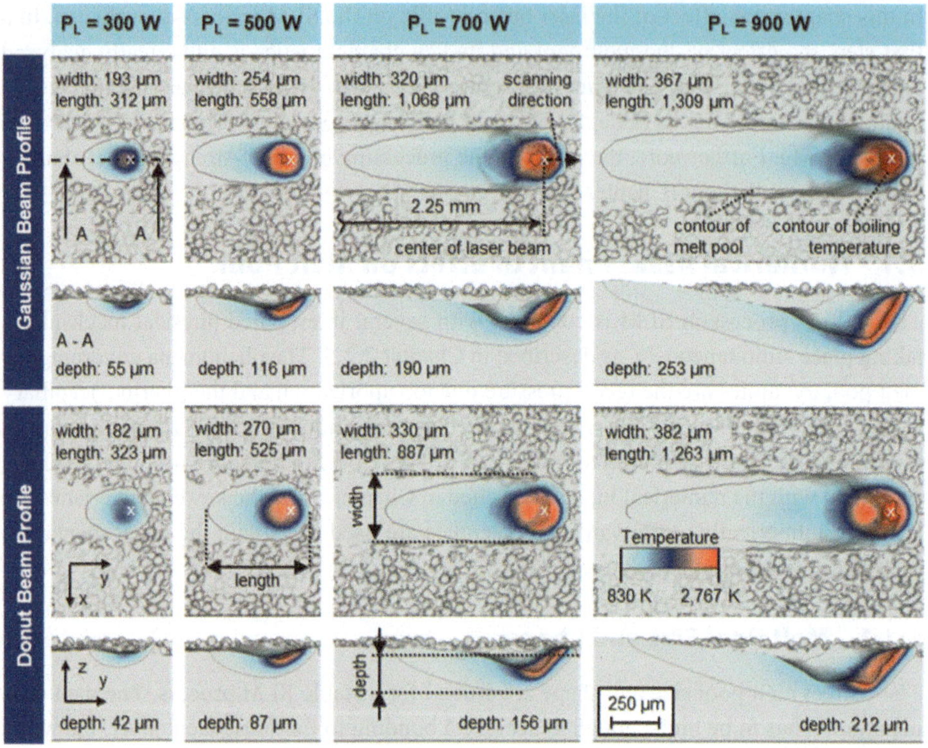

Figure 5-1: Melt pool shapes caused by the Gaussian and donut beam profile at scanning speed of 1,000 mm/s for selected laser powers at the position of y=2.25 mm in the scanning direction (numerical results).

To quantitatively describe the differences in the melt pool dimensions, Figure 5-2 gives an overview of the melt pool width, depth, length, and volume for different process parameters and both beam profiles. The melt pool width is ~110 µm at an LED of 0.15 J/mm and increases logarithmically to ~360 µm at an LED of 0.9 J/mm for the Gaussian beam profile. For the donut beam profile, the melt pool width increases logarithmically from ~100 µm to ~380 µm. Therefore, the melt pool width is slightly lower at low LEDs and slightly higher at high LEDs for the donut beam profile. The melt pool depth is approximately 25 µm at an LED of 0.15 J/mm for both beam profiles and increases linearly with an increase in LED to a depth of ~250 µm for the Gaussian and to ~210 µm for the donut beam profile at an LED of 0.9 J/mm. As a result, the melt pool depth is on average 20% higher for the Gaussian beam profile and the absolute difference increases with an increase in LED. Furthermore, the melt pool length increases linearly with an increase in laser power for both beam profiles. While the difference between the melt pool length produced by both beam profiles is negligible at a laser power of $P_L = 300$ W, it gets more significant at higher laser powers, especially at low scanning speeds. Additionally, the scanning speed does not have a noticeable effect on the melt pool length at low laser powers. Moreover, the melt pool volume produced by both beam profiles is very similar for low LEDs but larger for the Gaussian beam profile at high LEDs. This is as a result of the differences in melt pool width, depth and length discussed before.

The higher melt pool depth produced by the Gaussian beam profile can be seen as an advantage as it potentially enables an increase in layer thickness. In contrast, the lower melt pool width at high LEDs is a disadvantage as it reduces the possible hatch distance. However, it has to be mentioned that other phenomena will play a crucial role in this context, such as the stability of the melting process, the occurrence of defects (e.g., porosity) as well as the actual cross-sectional shape of the melt pool that determines the necessary overlap and layer thickness.

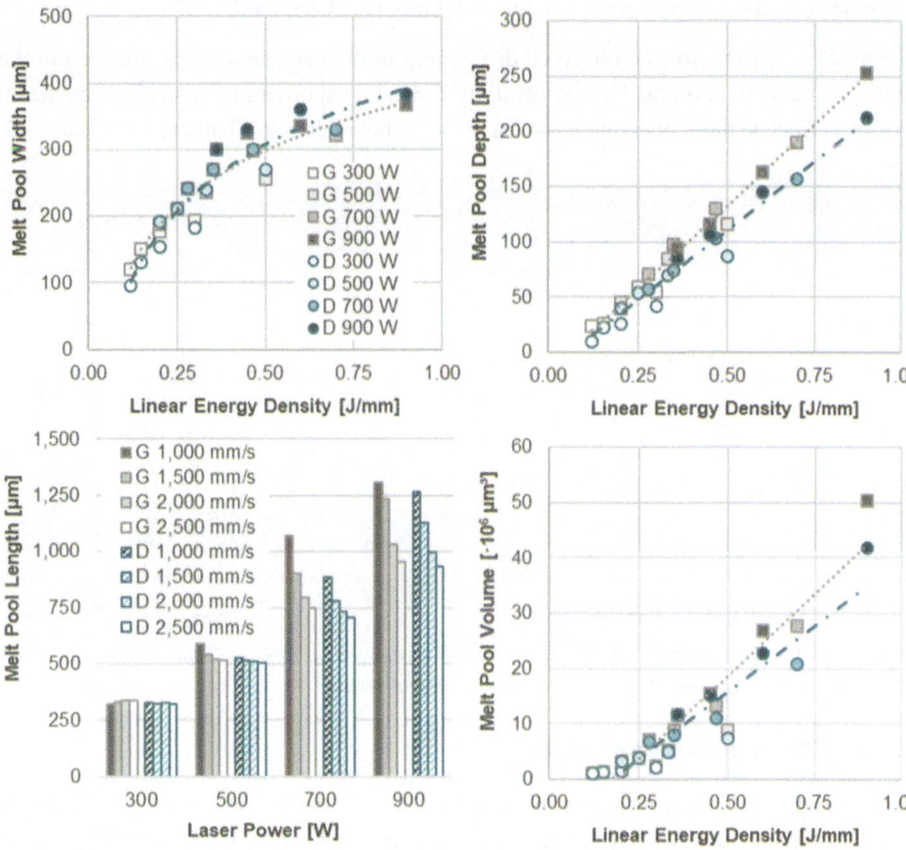

Figure 5-2: Melt pool width, depth, length and volume at different process parameters for the Gaussian (G) and donut (D) beam profile averaged over different positions along the scan track (numerical results).

In addition to the above, Figure 5-3 provides an overview of the width-to-depth (*w/d*) and length-to-width (*l/d*) ratio for different process parameters. The melt pool width-to-depth ratio varies between 1.7 and 9.5, decreases exponentially with an increase in LED and is smaller for the Gaussian compared to the donut beam profile for all process parameters (on average ~20%). Therefore, the melting process operates in heat conduction mode

($w/d > 3$) and transition mode ($3 > w/d > 2$) for almost all process parameters. Keyhole welding ($w/d < 2$) is initiated above LEDs of ~0.65 J/mm for the Gaussian and above ~0.75 J/mm for the donut beam profile for single tracks. Furthermore, the melt pool length-to-width ratio varies between 1.5 and 3.3, increases with an increase in scanning speed and is, on average, more than 10% higher for the Gaussian beam profile.

The circumstance that keyhole welding starts at higher LEDs and the lower length-to-width ratio can be seen as advantage of the donut beam profile. The higher threshold-energy for keyhole welding potentially increases the maximum energy input that still allows a stable melting process, and due to the shorter melt pools, the balling tendency is decreased based on the Plateau-Rayleigh instability (see Chapter 2.2.5).

The primary reasons for the observed differences in the melt pool shape are seen in the beam profile's effect on the thermo-capillary flow and vaporization. Therefore, the beam profile's effect on these phenomena is discussed in more detail in Chapters 5.1.2 and 5.1.3.

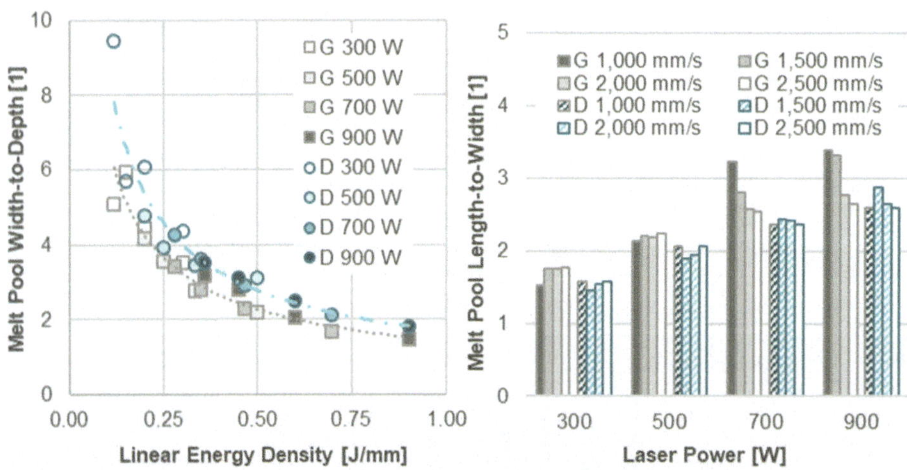

Figure 5-3: Melt pool width-to-depth and length-to-width ratio at different process parameters for the Gaussian (G) and donut (D) beam profile (numerical results).

5.1.2 Thermo-Capillary Flow

Thermo-capillary flow (or Marangoni-flow) is a surface driven flow that is based on the temperature dependency of the surface tension (see Chapter 2.2.5) [11, 98, 254]. Several studies underpin the importance of the thermo-capillary flow for the overall melt pool dynamics in SLM [22, 98]. Positive effects of a strong thermo-capillary flow are the enhancement of heat dissipation in the melt pool and potential increase in the melt pool width. In contrast, it also enhances the risk of gas-entrapment and spatters ejected from the melt pool due to high surface melt velocities [37, 98, 194].

To investigate the effect the laser beam profile has on the thermo-capillary flow, the surface temperature, the surface tension gradients and the melt velocities are analyzed. As the surface tension is directly related to the temperature (see Table 3-1 on page 47), the thermo-capillary flow will be stronger, the higher the temperature gradients in the melt pool are [176, 244]. Figure 5-4 exemplarily shows the melt pool surface temperatures for different process parameters for the Gaussian and donut beam profile. It can be seen that there is a visible effect of the beam profile on the surface temperature. The maximum temperature varies between 1,554 K and 4,526 K and increases with an increase in laser power. Overall, the maximum temperatures are higher for the Gaussian beam profile and exceed the boiling temperature (T_b = 2,767 K) in most cases. This is possible as the saturation temperature increases e.g., when vapor pressure acts on the surface [134, 189]. Furthermore, the surface area with temperatures exceeding the boiling temperature is larger for the Gaussian beam profile, indicating a more significant vaporization (see Chapter 5.1.3).

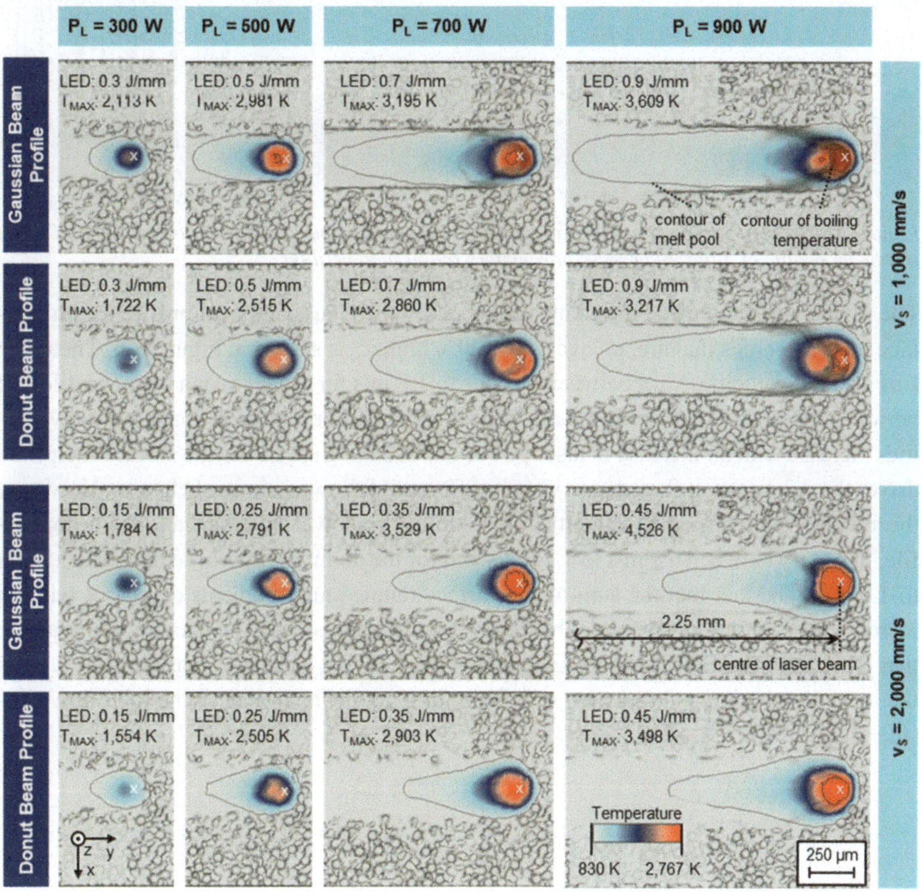

Figure 5-4: Melt pool surface temperatures caused by the Gaussian and donut beam profile at different process parameters and position y=2.25 mm in the scanning direction (numerical results).

The fact that the maximum temperatures are higher for the Gaussian beam profile indicates higher temperature gradients and, as a consequence, higher surface tension gradients. This is confirmed by Figure 5-5, which displays the total of the surface tension gradients of all melt pool surface cells for different process parameters (see Chapter 3.3 for details on the calculation of the surface tension gradients). The total of the surface tension gradients varies between ~150 kN/m² and 1,250 kN/m² and increases with an increase in LED. On average, this value is ~20% higher for the Gaussian beam profile due to the higher maximum temperatures.

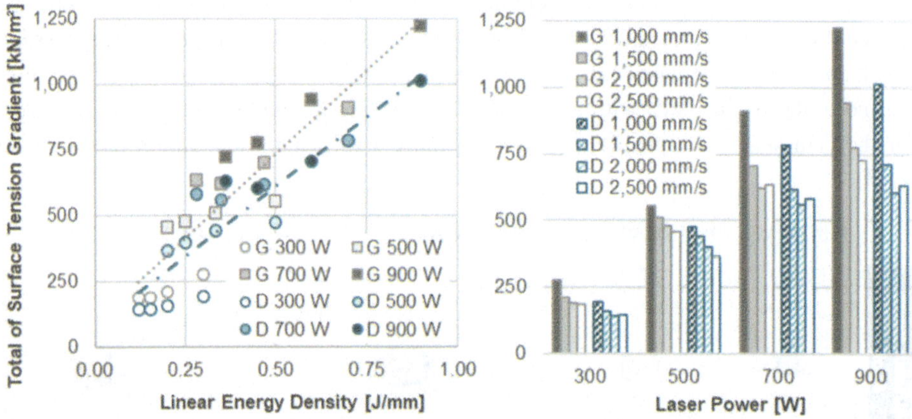

Figure 5-5: Total of surface tension gradients of the melt pool caused by the Gaussian (G) and donut (D) beam profile at different process parameters (numerical results).

The higher surface tension gradients produced by the Gaussian beam profile lead to surface melt velocities that are, on average, ~20% higher, as shown in Figure 5-6. The mean surface melt velocities vary between 1.3 m/s to 1.8 m/s for the Gaussian beam profile and between 1.0 m/s to 1.5 m/s for the donut beam profile. For both beam profiles, the velocities increase with an increase in laser power and scanning speed. As the surface melt flow drives the volume melt flow due to viscous effects, it follows a similar behavior. However, the average volumetric melt velocities are significantly lower. For the Gaussian beam profile, they vary between 0.7 m/s to 1.15 m/s and for the donut beam profile between 0.5 m/s to 1.0 m/s. The fact that the sum of the surface temperature gradients decreases with an increase in scanning speed, while the melt velocities increase simultaneously demonstrate that other phenomena that are not analyzed in this context, such as vaporization as well as the local distribution and direction of temperature gradients, influence the melt flow considerably.

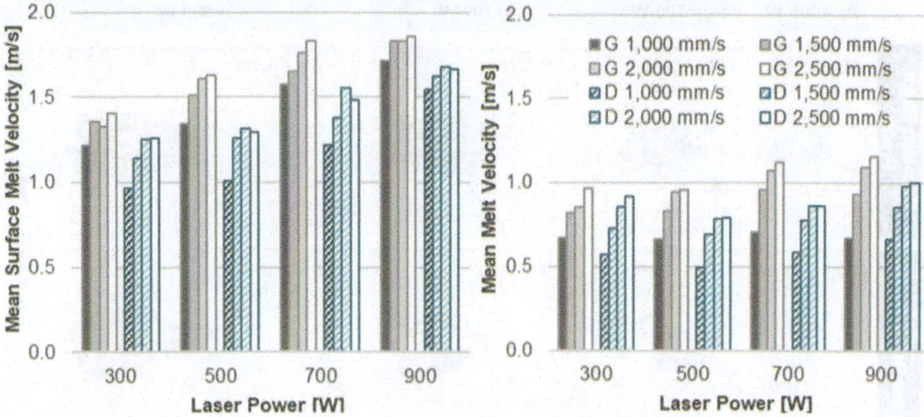

Figure 5-6: Mean surface- and volume-melt velocities caused by the Gaussian (G) and donut (D) beam profile at different process parameters (numerical results).

To investigate the direction of the melt flow, Figure 5-7 exemplifies the surface melt flow for selected process parameters and both beam profiles. The figure illustrates that the most significant melt flow occurs in the longitudinal direction (y-direction) behind the laser-interaction zone and in the reverse direction of the laser scanning direction (negative y-direction). Additionally, high melt velocities partly occur at the front edge of the melt pool when single powder particles are molten. This is a common phenomenon for spatter generation [31, 101]. The radial melt flow (x-direction) is considerably lower than the longitudinal one. This is shown in Figure 5-8, which displays the volumetric melt velocities divided into radial, longitudinal and vertical (z-direction) direction. In the radial and vertical direction, the average melt velocities vary between 0.15 – 0.3 m/s for both beam profiles, increase with scanning speed and show a small decrease with an increase in laser power. The average melt velocities in the longitudinal direction are significantly higher and vary between 0.45 – 1.0 m/s for the Gaussian and 0.35 – 0.8 m/s for the donut beam profile and increase with an increase in laser power as well as scanning speed. Therefore, the effect of the beam profile on the melt velocities is considerably high in the longitudinal direction, while it is negligible in the radial and vertical directions.

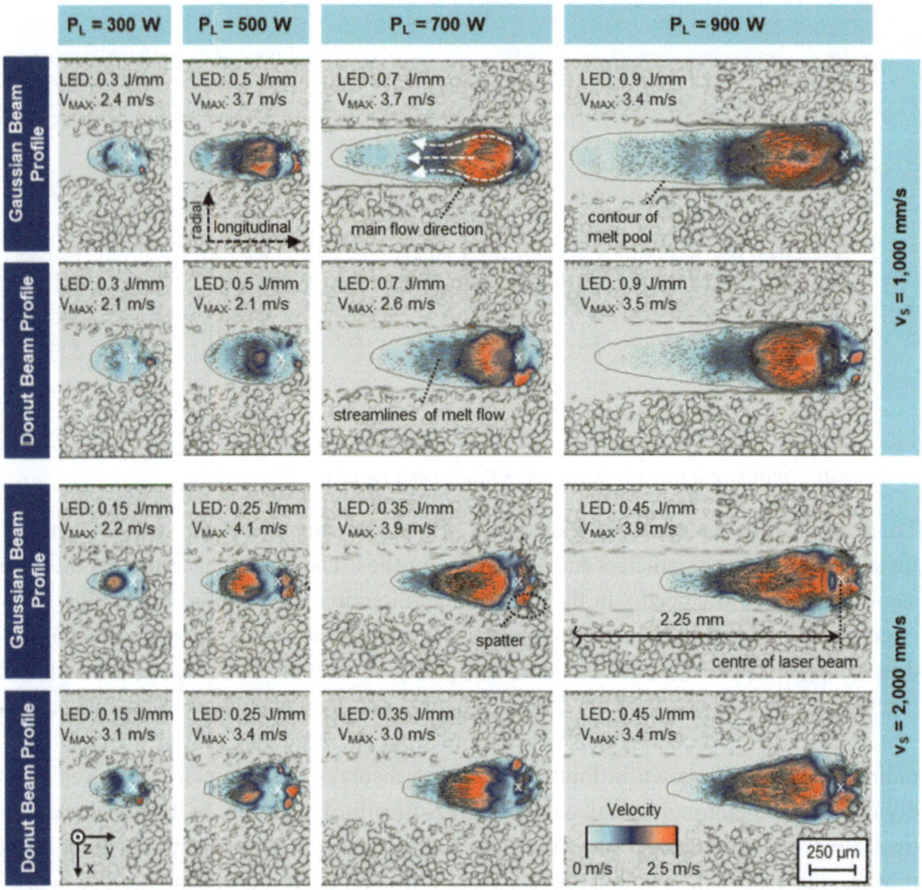

Figure 5-7: Melt pool surface velocities (color) and flow direction (streamlines) caused by the Gaussian and donut beam profile at different process parameters and position y=2.25 mm in scanning direction (numerical results).

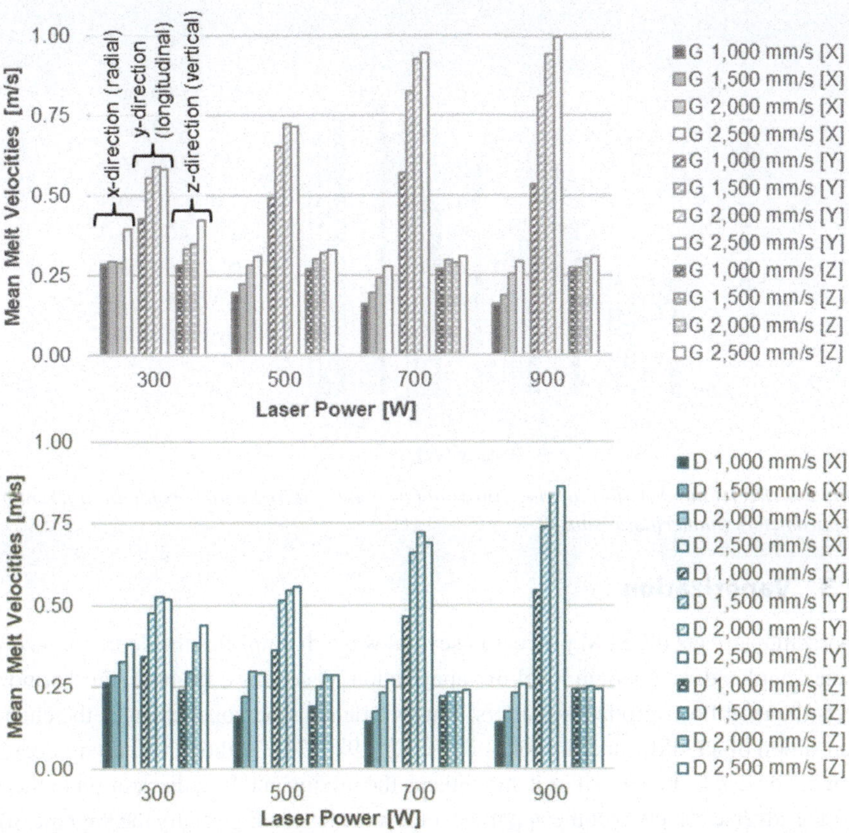

Figure 5-8: Radial (X), longitudinal (Y) and vertical (Z) volume melt velocities caused by the Gaussian (G) and donut (D) beam profile at different process parameters (numerical results).

To quantitatively describe the importance of the thermo-capillary flow for the overall heat dissipation, the Peclet number *Pe* is calculated that defines the ratio of convective heat transfer to conductive heat transfer (see Chapter 2.2.5 and Equation 2.11 on page 25). Figure 5-9 shows the Peclet numbers for different process parameters and both beam profiles, assuming an average density of $\rho = 2{,}400$ kg/m^3, a specific heat capacity of $C_P =$ 1,040 J/(kgK) and a thermal conductivity of $k = 100$ W/(m·K) (see Figure 3-5 and Figure 3-6 on page 48). The Peclet number varies between 5 and 17, increases with an increase in laser power and is on average 20% higher for the Gaussian compared to the donut beam profile. The results confirm that the convective heat transfer (thermo-capillary flow) is the primary mechanism driving the heat dissipation in the melt pool. This agrees with other studies that reported Peclet numbers between 5 – 100 [62, 119, 194].

On the one hand, the stronger thermo-capillary flow can be seen as an advantage of the Gaussian beam profile as it amplifies the heat dissipation in the melt pool. On the other hand, the higher melt velocities increase the risk of spatter ejection and gas-entrapment in the melt pool that can cause porosity.

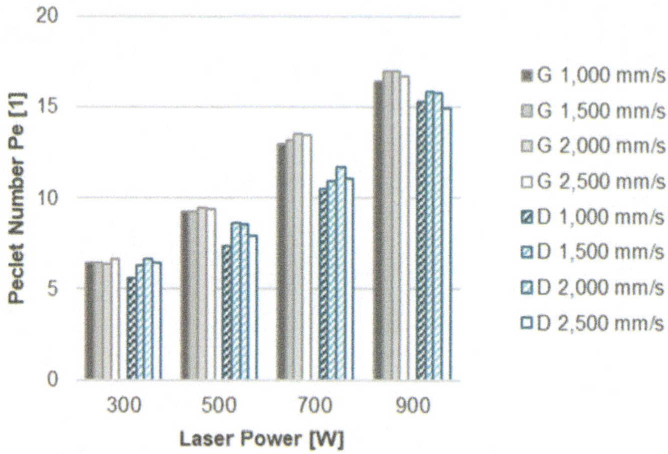

Figure 5-9: Peclet number (Pe) for the Gaussian (G) and donut (D) beam profile at different process parameters (numerical results).

5.1.3 Vaporization

Vaporization affects the SLM process in several ways: It amplifies the depression zone in the melt pool and, at a certain level of vaporization, generates a keyhole. Furthermore, it affects the melt flow, produces spatters, causes a denudation zone, changes the chemical composition of an alloy, and provokes defects [1, 98, 132]. Although a certain degree of vaporization can be beneficial as it strengthens the mixing and heat dissipation in the melt pool, the adverse effects become apparent at elevated levels. Especially the keyhole effect, which occurs at a certain threshold of vaporization pressure deteriorates the stability of the process and is a limiting factor for the maximum feasible energy input. Therefore, a reduction in vaporization can, in general, be seen as a positive effect that potentially increases the stability and productivity of the SLM process.

To investigate the beam profile's effect on the vaporization, the evaporation pressure is calculated numerically for different process parameters. Figure 5-10 exemplarily illustrates the vapor pressure acting on the melt pool surface for different process parameters and both beam profiles. It demonstrates that the vapor pressure acts at the front edge of the melt pool in the vicinity of the highest laser intensity (the center of the laser beam). Due to changes in the melt pool shape, multiple reflections and surface melt flow, several vapor pressure spots can occur as shown e.g., for the Gaussian beam profile at a laser power of P_L = 900 W and a scanning speed of v_S = 1,000 mm/s. Furthermore, the beam profile has a noticeable effect on the vapor pressure as it is considerably higher for the Gaussian beam profile for all parameters shown. Figure 5-11 quantitatively compares the vapor pressure force generated by the Gaussian and donut beam profile at different process parameters. The total vapor pressure force F_{VAP} varies between 0 - 4 nN and increases with laser power, while the effect of the scanning speed is comparatively low. Vaporization starts at P_L = 500 W for the Gaussian beam profile and at P_L = 600 W for the donut beam profile and the vapor pressure force is, on average, about 1 nN smaller for the donut beam profile at the same process parameters. Consequently, the vapor pressure produced by the donut beam profile at a specific laser power is about the same vapor pressure produced by the Gaussian beam profile at a laser power that is 100 W lower. This indicates that the threshold laser power at which keyhole welding occurs can be increased by ~100 W with the donut beam profile for single tracks, based on the assumption that keyhole welding is mainly driven by the total vapor pressure acting on the melt pool surface. This concurs with the results shown in Figure 5-3 (page 64) that visualizes the width-to-depth ratio of the melt pool at different process parameters. Based on the indicated trend lines, keyhole welding ($w/d \leq 2$) occurs at LEDs $\geq$ 0.75 J/mm (e.g., P_L = 750 W and v_S= 1,000 mm/s) for the donut beam profile and at LEDs $\geq$ 0.65 J/mm (e.g., P_L = 650 W and v_S = 1,000 mm/s) for the Gaussian beam profile.

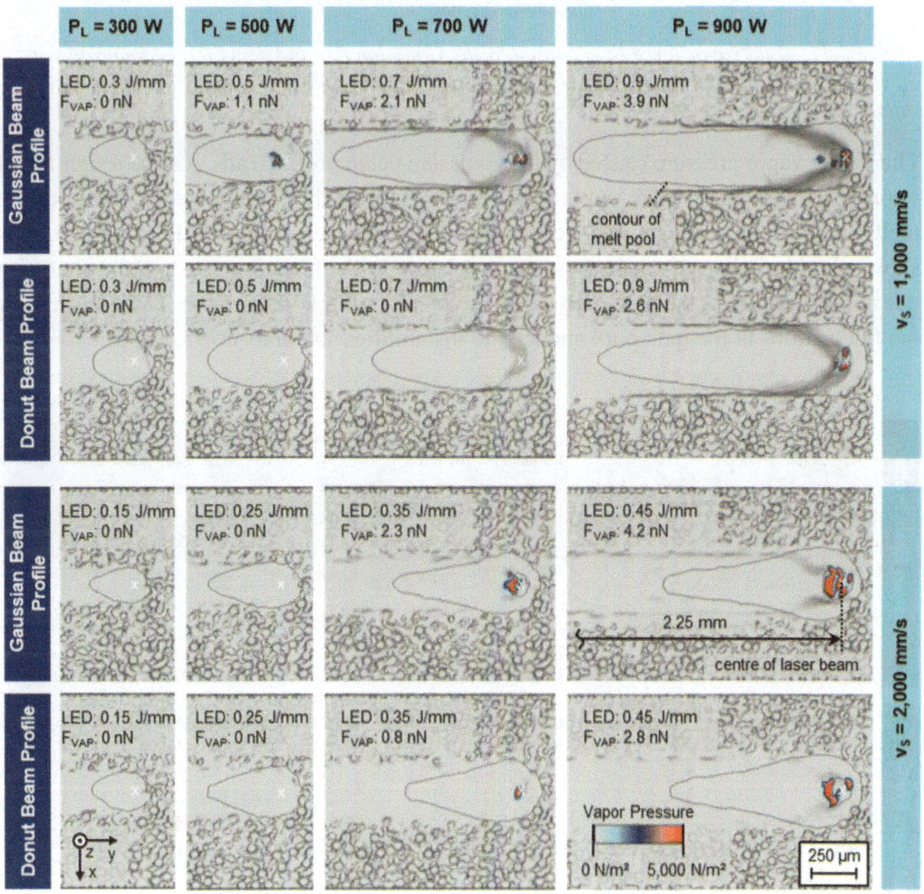

Figure 5-10: Vapor pressure caused by the Gaussian and donut beam profile at different process parameters and position y=2.25 mm in the scanning direction (numerical results).

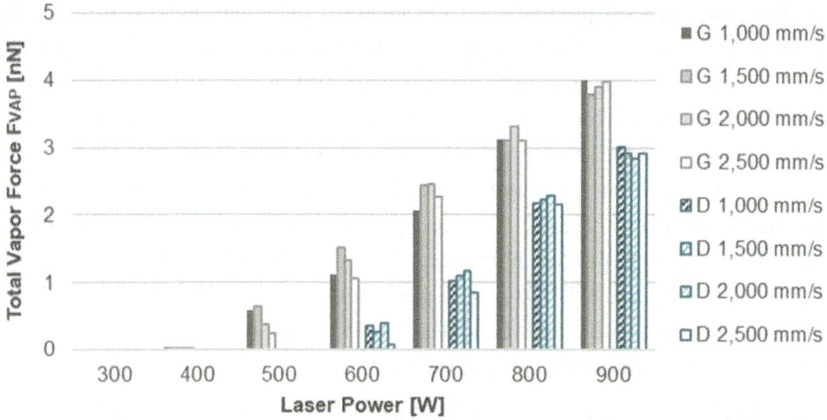

Figure 5-11: Total vapor force F_{VAP} caused by the Gaussian (G) and donut (D) beam profile at different process parameters averaged over different positions along the scan track (numerical results).

The higher vapor pressure caused by the Gaussian beam profile leads to a more significant impression zone in the melt pool, as shown in Figure 5-12. Consequently, the resulting melt pools are on average 20% deeper (see Figure 5-2 on page 63). Furthermore, it is shown that the origin of vapor pressure in relation to the laser center is rapidly changing due to the variance of the melt pool shape as well as multiple-reflections. This can amplify the fluctuations in the melt pool and lead to instabilities and defects [98, 181].

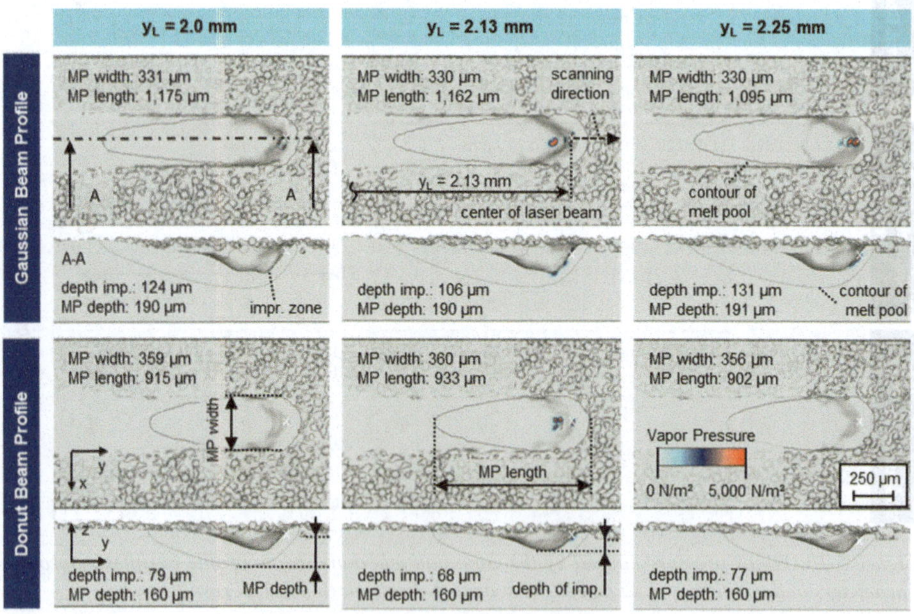

Figure 5-12: Melt pool geometry and vapor pressure caused by the Gaussian and donut beam profile at a laser power of P_L = 700 W and scanning speed of v_S = 1,000 mm/s at different positions along the scan track (numerical results).

To quantitatively describe the fluctuations in the impression zone, Figure 5-13 displays the relative depth of the impression for different positions along the scan track for a laser power of $P_L = 700$ W and different scanning speeds. The variation in the depth of the impression zone follows a random pattern and is significantly higher for the Gaussian beam profile. This is confirmed by the data in the graph visualizing the total fluctuation for different process parameters in Figure 5-13. In this context, the total fluctuation is defined as the sum of all changes in the depth of the impression zone from $y = 1.25$ mm to $y = 2.5$ mm along the scan track. The total fluctuation varies between ~25 µm to ~50 µm for laser powers of $P_L = 300 - 500$ W at which only limited vaporization occurs (see Figure 5-11) and increases up to 125 µm for laser powers of $P_L = 700 - 900$ W. To verify that this fluctuation is a result of vaporization, Figure 5-14 compares the fluctuations in the impression zone caused by the Gaussian beam profile to numerical results obtained with a disabled vaporization pressure. The outcome demonstrates that the fluctuations are reduced by 50% in this specific case. This confirms that the vapor pressure is the primary effect causing the melt pool fluctuations.

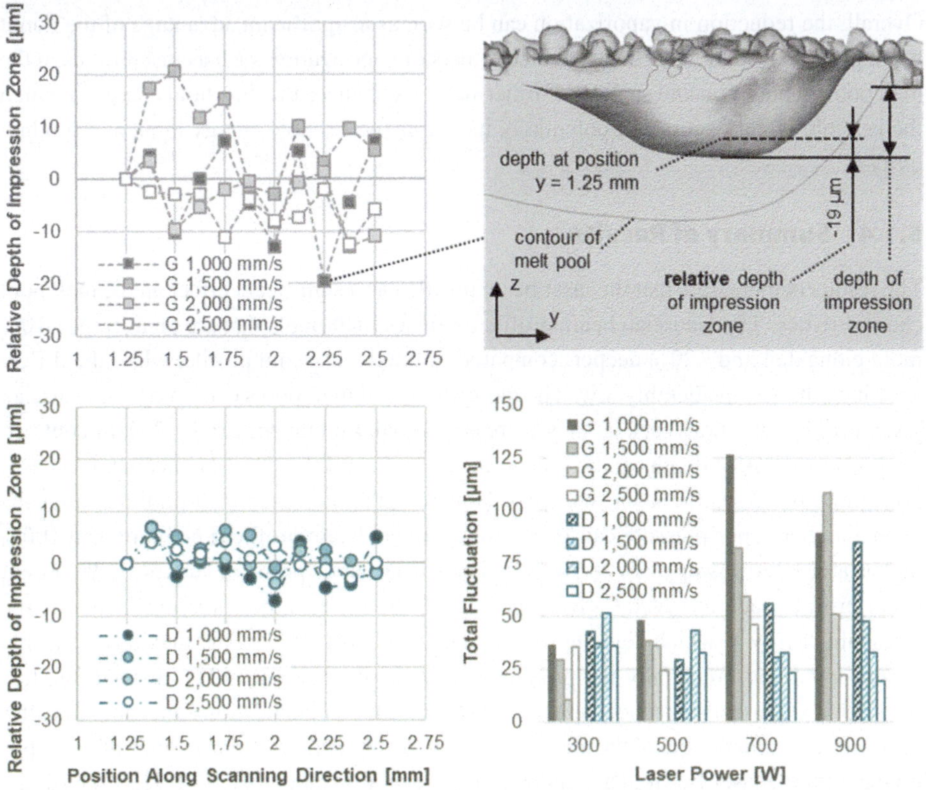

Figure 5-13: Depth of the impression zone caused by the Gaussian (G) and donut (D) beam profile at different positions along the scan track for a laser power of $P_L = 700$ W as well as a comparison of total fluctuation for different process parameters (numerical results).

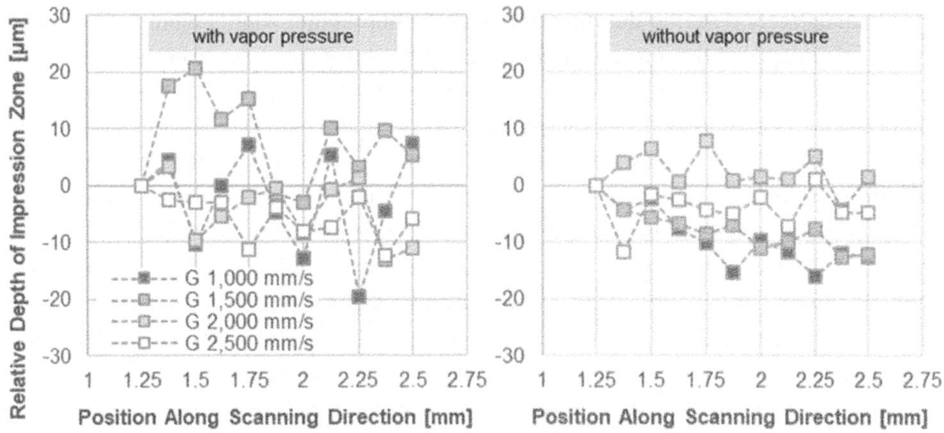

Figure 5-14: Depth of the impression zone caused by the Gaussian (G) beam profile at different positions along the scan track for a laser power of P_L = 700 W for simulations with and without vapor pressure (numerical results).

Overall, the reduction in vaporization can be seen as a significant advantage of the donut beam profile, as it attenuates the adverse effects of vaporization such as instabilities in the melt pool, denudation and spatters. Furthermore, it enhances the threshold energy at which the keyhole effect occurs and potentially increases the feasible energy input and productivity.

5.1.4 Summary of Results

The results demonstrate that the laser beam profile has a significant effect on the melt pool characteristics. The Gaussian beam profile produces melt pools that are on average ~10% more elongated and ~20% deeper, compared to the donut beam profile, while the difference in width is considerably low. The more elongated melt pools caused by the Gaussian beam profile are a drawback as they increase the balling tendency [98, 139]. In contrast, the higher melt pool depth can be seen as an advantage of the Gaussian beam profile as it potentially increases the layer thickness. However, the increased melt pool depth results from a higher vapor pressure caused by the Gaussian beam profile (i.e., more than 100% higher at a laser power of P_L = 700 W). The lower vapor pressure caused by the donut beam profile leads to several advantages. First, the lower vapor pressure increases the threshold energy at which keyhole welding occurs. This increases the maximum feasible energy input and potentially the productivity of the process [20, 137]. Second, the less significant vaporization reduces spatters, leads to a smaller denudation zone, does not alter the chemical composition of the alloy as much, and attenuates the interference of the vapor plume with the laser beam that can deteriorate laser absorption [77, 132, 137]. Third, the lower vapor pressure decreases the fluctuations in the melt pool. This indicates a stabler melt pool that will contribute to more homogeneous melt tracks [25, 131]. In addition to the above, the Gaussian beam profile causes higher temperature gradients in the melt pool, leading to a stronger thermo-capillary flow, especially in the longitudinal direction. As a result, the melt velocities are, on average, 20% higher than the melt velocities caused by the donut beam profile. Although an increase in melt velocity enhances the heat dissipation, it also increases instability in the melt pool and ejection of spatters [22, 69, 181].

5.2 Effect on Single Tracks of AlSi10Mg

Single track experiments are performed to examine the laser beam profile's effect on the SLM process experimentally. Therefore, the single tracks are categorized by appearance (Chapter 5.2.1) and their cross-sectional size and shape (Chapter 5.2.2). Furthermore, the results are summarized (Chapter 6.1.3).

Single track experiments are a simple method to evaluate the basic characteristics of the melting process in SLM [96, 245]. Studies have shown that a continuous and smooth melt track is the basis for a stable multi-layer SLM process [30, 96]. Nevertheless, it must be said that there are limitations in the direct transfer of the results to the multi-layer SLM process. This is because some essential phenomena that are relevant for the multi-layer SLM process are neglected, such as hatch distance, pre-heating due to neighboring tracks and denudation [3, 205].

5.2.1 Appearance

The appearance of single tracks in the literature is categorized into three to six types, depending on the grading [30, 96, 245]. In this thesis, three types are defined as shown in Figure 5-15 since a more detailed grading is not considered relevant for the analysis:

- *Type I – Continuous melt track*: The melt track topology is homogeneous and continuous. It can be defined as the desired shape for the single tracks [96, 245].
- *Type II – Occasionally broken melt track*: The melt tracks have short discontinuities in the topology. They can come from an energy input that is too low, long melt pools that start to initiate balling or instabilities induced by vaporization or high melt velocities [30, 245].
- *Type III – Discontinuous melt track*: The single tracks are characterized by discontinuities and single droplets along the path. They are a result of an energy input that is too low. Single segments and droplets are formed due to the limited penetration of the melt pool into the subsequent layer (see Plateau-Rayleigh instability explained in Chapter 2.2.5) [22, 30, 139].

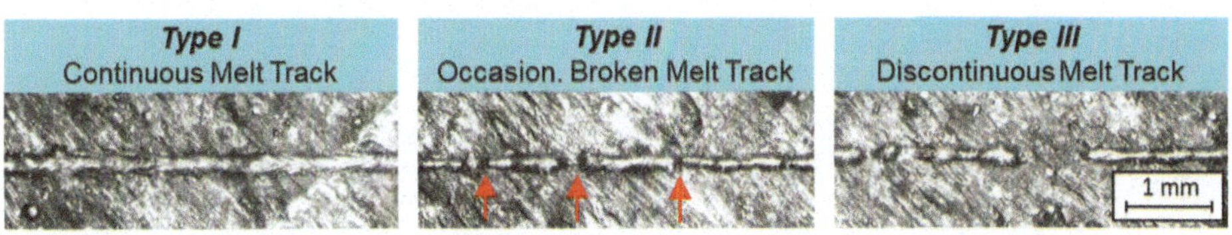

Figure 5-15: Different types of single tracks, based on their appearance (top view).

Table 5-1 exemplarily shows top-view pictures of the single tracks for both beam profiles produced at different process parameters. Furthermore, Figure 5-16 allocates the single tracks to the three categories shown in Figure 5-15. It must be mentioned that the allocation of the tracks to the categories is not always distinct and are subjective to a certain degree. Despite this, some general insights can be taken from the figures:

- Discontinuous melt tracks (type III) occur at low energy inputs and homogenous melt tracks (type I) at high energy inputs.
  - For the Gaussian beam profile, discontinuous melt tracks (type III) occur for laser powers of $P_L \leq 600$ W and LEDs ≤ 0.4 J/mm and homogeneous melt tracks (type I) at a laser power of $P_L = 700$ W and scanning speed of $v_S = 1{,}000$ mm/s.
  - For the donut beam profile, discontinuous melt tracks (type III) occur for laser powers of $P_L \leq 500$ W and LEDs ≤ 0.26 J/mm and homogeneous melt tracks (type I) at LEDs ≥ 0.46 J/mm.
- The single tracks produced by the donut beam profile are overall more continuous and show fewer irregularities as well as discontinuities. As a result, the region of homogeneous (type I) and occasionally broken melt tracks (type II) is larger for the donut beam profile.
- The region of homogenous and occasionally broken melt track (types I and II) exceed the limits of the process parameters tested. Once can expect that they can be produced with higher laser powers and eventually higher scanning speeds than tested for both beam profiles[3].

The circumstance that the region of melt tracks of types I and II is larger for the donut beam profile indicates a stabler melting process. The irregularities observed in the type I and type II melt tracks can come from the Plateau-Rayleigh instability, irregularities in the powder bed, or instabilities in the melt pool induced by vaporization or high melt velocities [25, 96]. The numerical results laid out in the previous chapter reveal that the surface melt velocities in the melt pool produced by the Gaussian beam profile are on average ~20% higher (see Figure 5-6) and that the total vapor pressure is on average 1 nN higher (equal to ~100% at a laser power of 700 W, see Figure 5-11). This could be an explanation for the more homogeneous melt tracks produced by the donut beam profile.

[3] It was not possible to test higher laser powers at the time of the experiments due to malfunctions of the laser.

Table 5-1: Top view pictures of single tracks produced with the Gaussian and donut beam profile for different process parameters.

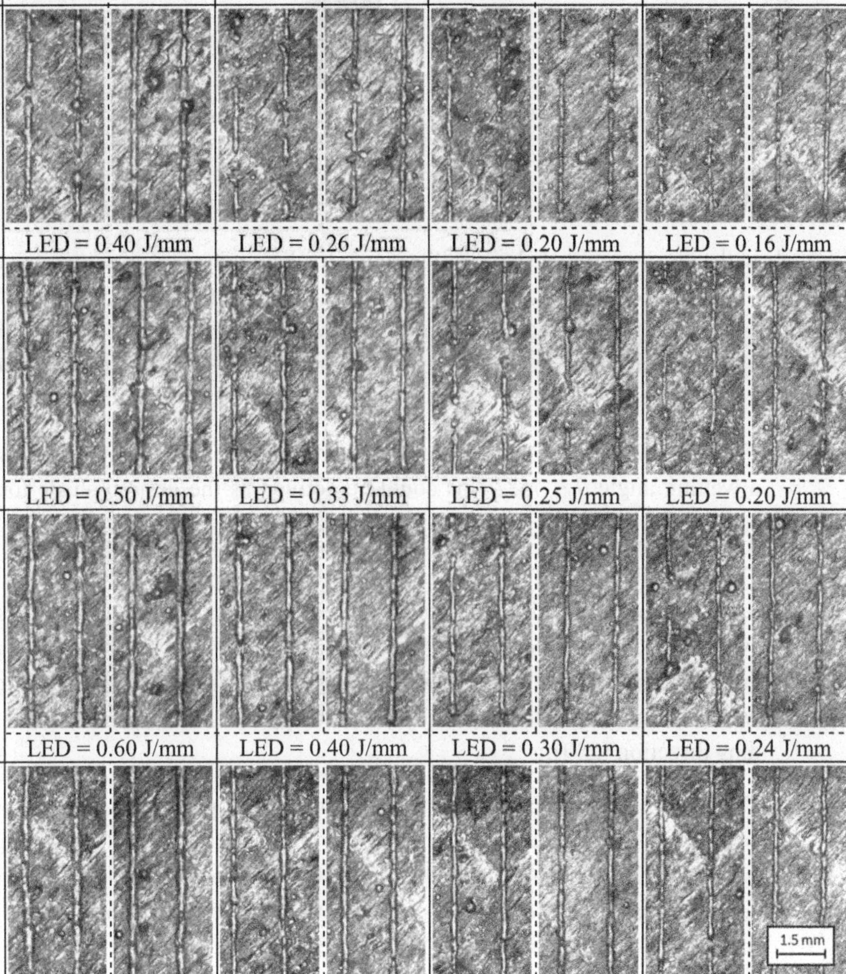

	Scanning Speed vs							
	1,000 mm/s		1,500 mm/s		2,000 mm/s		2,500 mm/s	
Laser Power P_L	Gaussian	Donut	Gaussian	Donut	Gaussian	Donut	Gaussian	Donut
400 W	LED = 0.40 J/mm		LED = 0.26 J/mm		LED = 0.20 J/mm		LED = 0.16 J/mm	
500 W	LED = 0.50 J/mm		LED = 0.33 J/mm		LED = 0.25 J/mm		LED = 0.20 J/mm	
600 W	LED = 0.60 J/mm		LED = 0.40 J/mm		LED = 0.30 J/mm		LED = 0.24 J/mm	
700 W	LED = 0.70 J/mm		LED = 0.46 J/mm		LED = 0.35 J/mm		LED = 0.28 J/mm	

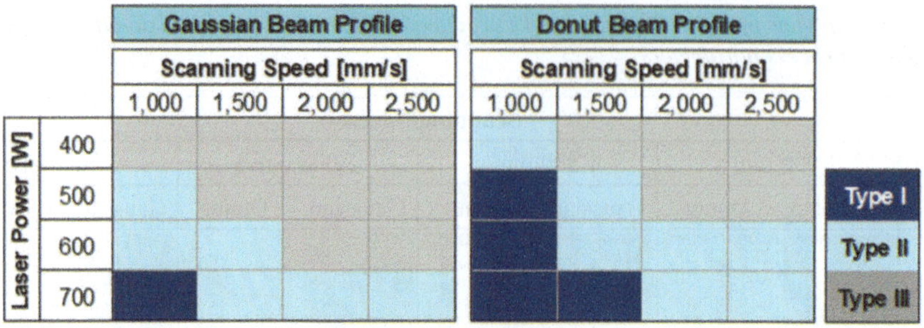

Figure 5-16: Categorization of single tracks produced with the Gaussian and donut beam profile based on top view appearance for different process parameters.

5.2.2 Width and Depth

To analyze the cross-sectional shape of the single tracks, micro-sections are prepared and surveyed. For each set of process parameters, five different tracks are analyzed and the results are averaged to take into account statistical variations of the process. Based on the micrographs, three different types of cross-sectional shapes are identified, as shown in Figure 5-17:

- *Type A – Symmetric Cross-Section*: The cross-section has a symmetric shape with a wetting angle $\theta < 75°$. This could be defined as the desired shape for cross-sections [3, 96, 248].
- *Type B – Protrusion*: The melt pool width in the base is greater than the width of the bulge and the melt pool penetrates deeper into the subsequent layers as for symmetric cross-sections (type A). Additionally, small impressions partially occur at the side of the bulge. This shape can result from vaporization combined with a strong longitudinal melt flow in the opposite direction of the laser-scanning direction [259].
- *Type C – Necking / Balling*: The melt pool barely penetrates the base and the cross-section of the bulge has an irregular or spherical shape. The wetting angle is $\theta > 75°$. This shape is the result of a low penetration of the melt pool into the subjacent layer. Consequently, single melt balls or segments are formed due to the Plateau-Rayleigh instability [69, 96, 234].

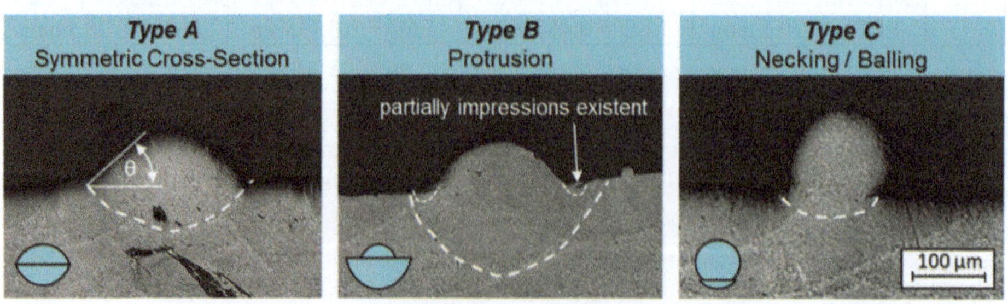

Figure 5-17: Types of single tracks depending on their cross-sectional shape.

Table 5-2 provides an overview of representative cross-sections for different process parameters and Figure 5-18 allocates the cross-sections to the categories shown in Figure 5-17. In the case that two different types of cross-sectional shapes are identified for a specific set of process parameters, both are indicated. Several insights can be taken from Figure 5-18:

- Cross-sections with a symmetrical shape (type A) occur for high LEDs and cross-sections with balling and necking (type C) occur for low LEDs.
    - For the Gaussian beam profile, symmetrical cross-sections (type A) can occur for LEDs ≥ 0.24 J/mm and cross-sections with balling and necking (type C) can occur for LEDs ≤ 0.4 J/mm.
    - For the donut beam profile, symmetrical cross-sections (type A) can occur for LEDs ≥ 0.20 J/mm and cross-sections with balling and necking (type C) can occur for LEDs ≤ 0.35 J/mm.
- The process window that consistently produces symmetric cross-section (type A) is larger for the donut beam profile.
- Cross-sections with protrusions (type B) occur at LEDs ≥ 0.45 J/mm and are found more often in the single tracks produced by the Gaussian beam profile.
- A strong correlation between discontinuous melt tracks (type III) and cross-section with necking or balling (type C) is observed. On the other hand, the correlation between continuous (type I) and occasionally broken (type II) melt tracks to symmetric cross-sections (type A) or cross-sections with protrusions (type B) is not as significant.
- Cross-sections with balling or necking (type C) are found in occasionally broken melt tracks (type II), but not in continuous melt tracks (type I).

Several studies indicate that that the process parameters producing symmetrical cross-sections foster a stable SLM process [96, 245]. Furthermore, protrusions stipulate significant vaporization and elongated melt pools that facilitate balling [3, 30]. Therefore, the fact that the process window that produces symmetrical cross-sections (type A) is larger and that protrusions are almost totally absent can be seen as an advantage of the donut beam profile.

Table 5-2: Representative cross-sectional shapes of single tracks manufactured with the Gaussian and donut beam profile for different process parameters

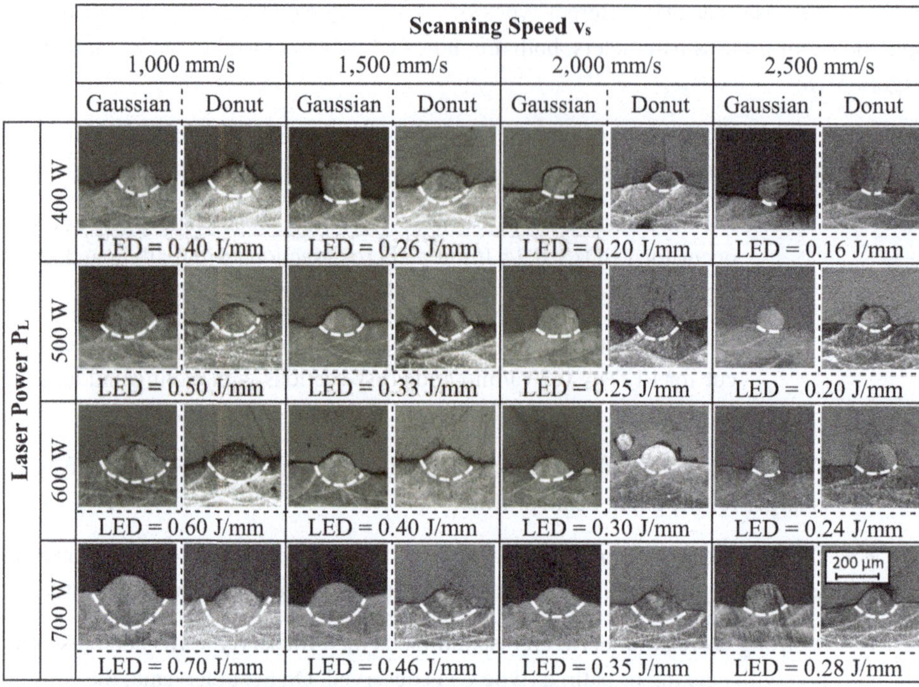

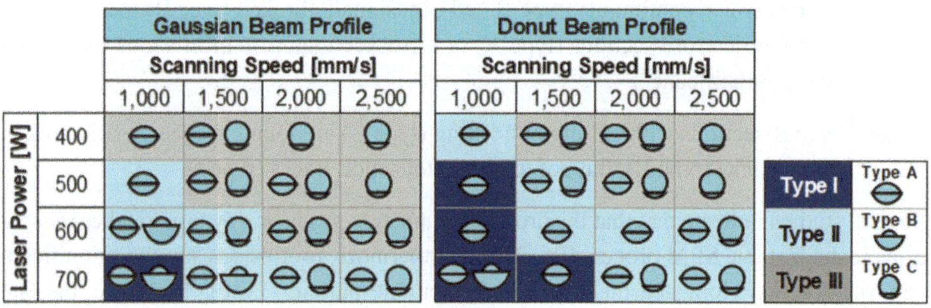

Figure 5-18: Categorization of single tracks manufactured with the Gaussian and donut beam profile depending on cross-sectional shape for different process parameters.

In addition to the appearance, the single melt tracks width and depth are analyzed (see Figure 5-19). It can be seen that the melt pool width increases logarithmically with an increase in laser power for both beam profiles and that the width is slightly higher for the Gaussian beam-profile at LEDs > 0.4 J/mm. The numerical results presented in Chapter 5.1.1 (see Figure 5-2 on page 63) show the same trends. However, the melt pool width is slightly higher for the donut beam in the simulation at LEDs > 0.4 J/mm. Considering the standard deviation in the experimental results, the difference between the experimental and numerical results is reasonable. In contrast, the melt pool depth shows a linear increase with an increase in LED for both beam profiles, while the gradient is higher for the Gaussian beam profile. Consequently, the melt pool depth produced by the Gaussian beam profile is noticeably higher for LEDs ≥ 0.4 J/mm (on average 30%), while it is very similar to the depth produced by the donut beam profile at LEDs < 0.4 J/mm. The results align with the numerical results presented in Chapter 5.1.1. To investigate if the differences in the melt pool width and depth caused by both beam profiles are statistically significant, a Wilcoxon-Signed-Rank-Test is performed (see Appendix A.6). The results reveal a statistically relevant effect of the beam profile on the melt pool depth for LEDs ≥ 0.4 J/mm with a confidence level of 95%, but not on the melt pool width.

Although the higher melt pool depth caused by the Gaussian beam profile can be seen as an advantage as it potentially increases the feasible layer thickness, several considerations suggest that it is a disadvantage. As explained in Chapter 5.1.3, the primary reason for the increased melt pool depth is the higher vapor pressure produced by the Gaussian beam profile at laser powers of $P_L \geq 500$ W. Vaporization is a critical disturbance to the SLM process that produces spatters, creates a denudation zone, can alter the chemical composition of the material and induces instabilities in the melt pool. Furthermore, vaporization is a limiting factor of the maximum feasible energy input, and therefore limits the productivity of the process.

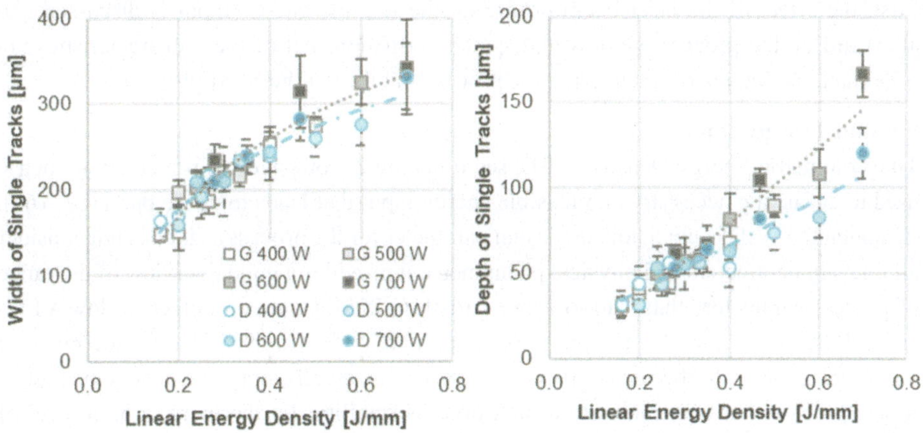

Figure 5-19: Melt pool width and depth of single tracks manufactured with the Gaussian (G) and donut (D) beam profile for different LEDs.

5.2.3 Summary of Results

For the analysis of the effect of the beam profile on single tracks, the appearance, as well as cross-sectional size and shape, were analyzed. The results show that the single tracks produced by the donut beam profile are overall more homogeneous and that the process window that produces the desired symmetric cross-sections is slightly larger for the donut beam profile. Furthermore, protrusions that are an indicator of significant vaporization and a strong longitudinal melt flow are avoided almost entirely by the donut beam profile. In addition, the beam profile has a relevant effect on the cross-sectional melt pool size and shape. While the melt pool width produced by both beam profiles is very similar, the melt pool depth caused by the Gaussian beam profile is, on average, 30% higher for LEDs $\geq$ 0.4 J/mm.

The results indicate a stabler melting process and a larger process window for the donut beam profile. The primary reasons for the observed differences are seen in the reduced vapor pressure and lower melt velocities caused by the donut beam profile that leads to a stabler melting process (see Chapter 5.1).

5.3 Effect on Multi-Layer SLM Process of AlSi10Mg

To investigate the effect of the laser beam profile on the multi-layer SLM process for AlSi10Mg, density cubes with different process parameters are manufactured, as explained in Chapter 4.1. At first, the effect of the beam profile on the process stability (Chapter 5.3.1) and afterward on the build rate is described (Chapter 5.3.2). Furthermore, a comparison to the results obtained for the single-track experiments is given (Chapter 5.3.3) and the results are summarized (Chapter 5.3.4).

5.3.1 Effect on Process Stability

First, the effect of the laser beam profile on the feasible energy input is discussed and afterward on the process window and spatter formation. All of these characteristics are important parameters characterizing the process stability of the SLM process.

Feasible Energy Input
The Volumetric Energy Density (VED, see Equation 2.3 on page 16) is a common metric used to define the necessary and feasible energy input that leads to dense parts [22, 164]. Although the VED is not a holistic design parameter for the process as it does not consider all relevant parameters and physical phenomena, it provides a good overview of the range of process parameters that lead to dense parts [16, 205]. Common defects at low VEDs are lack-of-fusion porosity and gas porosity at high VEDs [1, 69, 255]. A wide range of VEDs that lead to highly dense parts can be seen as an advantage as it demonstrates a robust process that contributes to a high process stability. Furthermore, a high level of feasible energy potentially increases the productivity of the process, as will be described in Chapter 5.3.2.

Figure 5-20 displays the density of all AlSi10Mg density cubes depending on different VEDs. The corresponding micro-sections are shown in Appendix A.7 for reference. It can be seen that the density of the cubes manufactured by the donut beam profile is, on average, higher than the ones manufactured by the Gaussian beam profile, especially at high VEDs. Above a VED of 80 J/mm^3 all cubes manufactured with the donut beam profile have a density > 99.5%, while most of the cubes manufactured with the Gaussian beam profile (7 out of 12) have densities lower than this value. For VEDs < 80 J/mm^3, the difference in the densities for the two beam profiles is not as significant at first sight but gets more noticeable when one focuses on the specimens manufactured with laser powers $P_L \geq$

600 W, as shown in Figure 5-21. In this case, the density of the cubes manufactured with a VED of 35-100 J/mm³ varies between 99-100% (region A) and the density is significantly lower at VEDs < 35 J/mm³ (region B) and VEDs > 100 J/mm³ (region C) for the Gaussian beam profile. For the donut beam profile, on the other hand, all cubes manufactured with a VED ≥ 35 J/mm³ have a density above 99.5% (region A) and significant porosity occurs at VEDs < 35 J/mm³ only (region B).

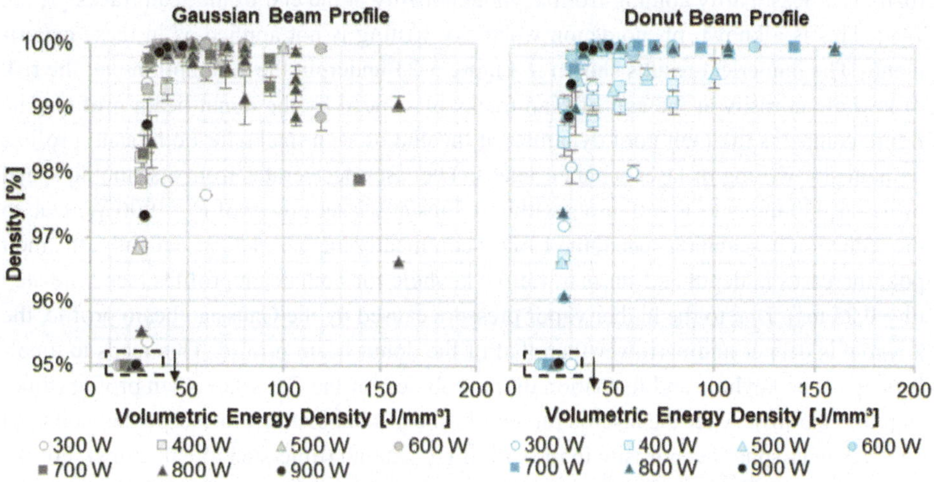

Figure 5-20: Density of specimens manufactured in AlSi10Mg for the Gaussian and donut beam profile for different VEDs and laser powers of P_L = 300 – 900 W.

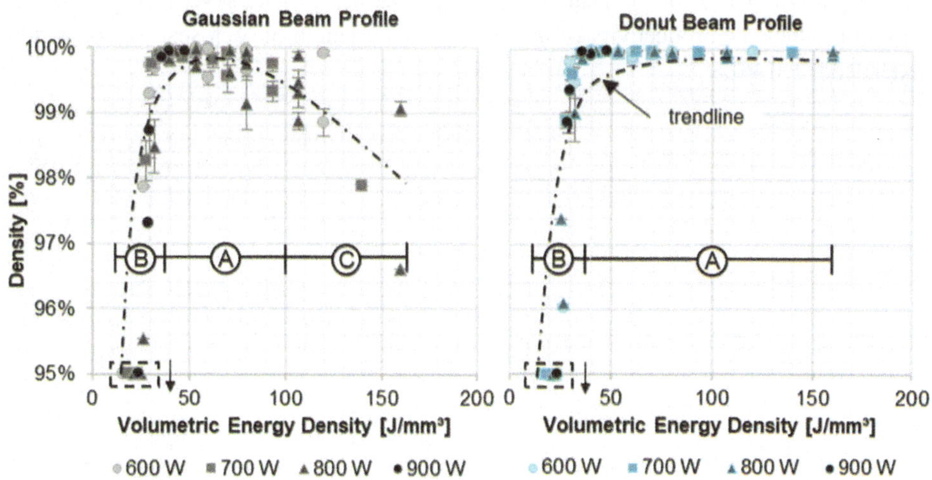

Figure 5-21: Density of specimens manufactured in AlSi10Mg for the Gaussian and donut beam profile for different VEDs and laser powers of P_L = 600 – 900 W. Trendlines indicate the course of density over VED (schematic).

To analyze the type and source of porosity, Figure 5-22 shows selected micro-sections of the specimens manufactured with a Gaussian beam profile for regions A, B and C. Micro-section 1 and 2 in Figure 5-22 show that the specimens manufactured with a

VED < 35 J/mm³ (region B) have a considerable amount of lack-of-fusion porosity, which is a result of an energy input that is too low (see Chapter 2.2.6 for details). For VEDs between 35 – 100 J/mm³ (region A), a low amount of spherical porosity is observed as shown in micro-section 3 of Figure 5-22. For VEDs > 100 J/mm³ (region C), a significant increase in spherical porosity with an increase in VED occurs and micro-section 4 (Figure 5-22) demonstrates that line-patterns can partly be identified. This indicates that the porosity is at least partly coming from keyhole porosity at the end of the scan tracks [1, 20, 141]. This is a known phenomenon when sky-writing is not applied as in these experiments. The numerical results shown in Figure 5-23 underpin the assumption that the risk of keyhole porosity at the end of scan tracks is reduced by the donut beam profile. The figure compares the melt pool dynamics at the end of scan tracks for both beam profiles without sky-writing or laser scanner delays. The laser decelerates from a scanning speed of v_S = 1,000 mm/s to v_S = 0 mm/s within 0.5 ms at a laser power of P_L = 900 W. Due to the reduction in scanning speed, the LED is continually increased. Consequently, the melt pool increases in depth and starts to form a keyhole for both beam profiles (see time-step t0 + 0.25 ms). Due to the higher vapor pressure caused by the Gaussian beam profile, the keyhole is deeper and narrower than that of the donut beam profile. This leads to a collapsing of the keyhole and formation of gas bubbles for the Gaussian beam profile (time-step t0 + 0.5 ms). After the laser is turned off, the gas bubbles fail to escape the melt pool and end up as spherical porosity in the solidified material (time-step t0 + 0.75 ms). For the donut beam profile, no collapsing and formation of gas bubbles occurs for the same process parameter. Based on the results, it is reasonable to assume that the amount of porosity can be partly reduced for the Gaussian beam profile by utilizing sky-writing. However, this will reduce the productivity of the process as the length of each scan vector will be extended [71, 146].

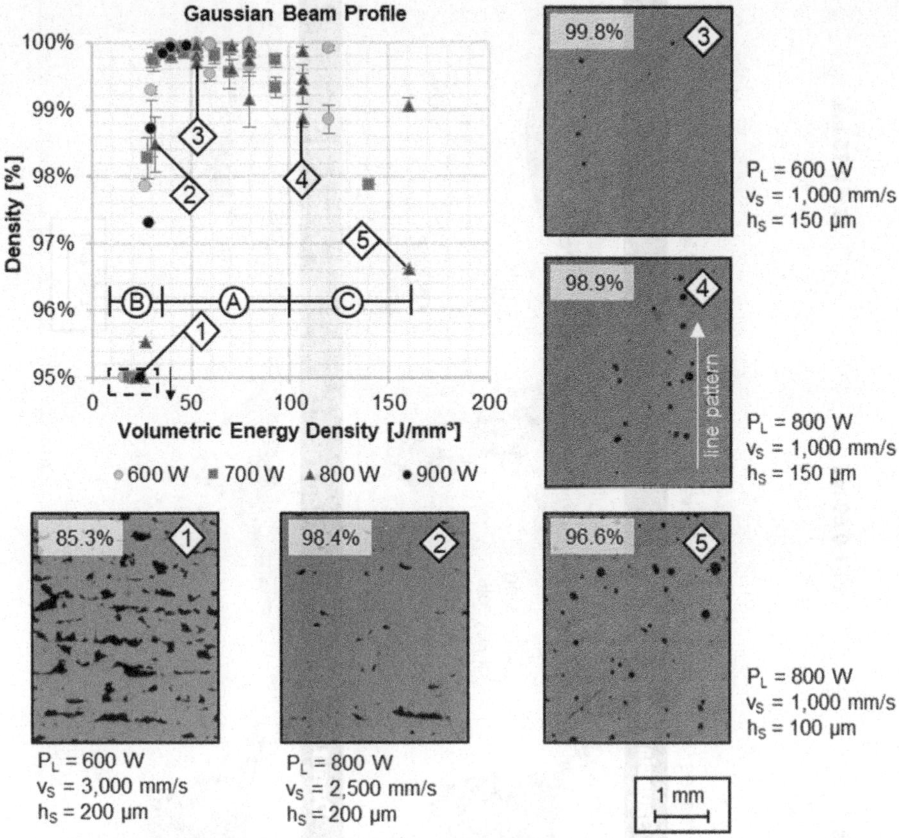

Figure 5-22: Representative micro-sections of specimens manufactured with the Gaussian beam profile in AlSi10Mg.

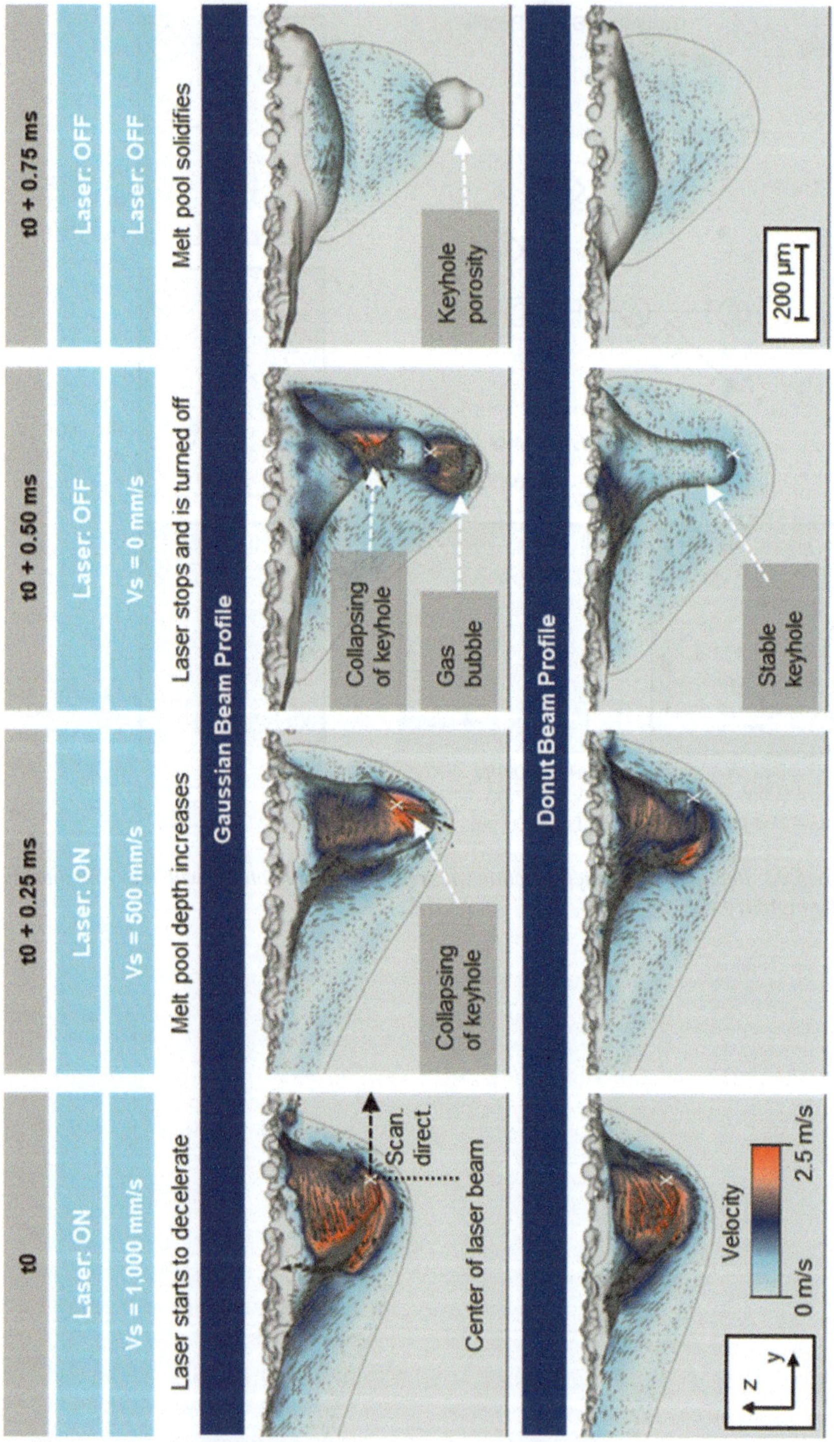

Figure 5-23: Melt pool dynamics at the end of scan tracks for a laser power of P_L = 900 W and scanning speed decelerating from v_S = 1,000 mm/s for the Gaussian and donut beam profile.

Figure 5-24 shows selected micro-sections of density cubes manufactured with the donut beam profile. As for the Gaussian beam profile, the density cubes with VEDs < 35 J/mm³ (region B) show significant lack-of-fusion porosity. For VEDs ≥ 35 J/mm³ (region A) the micro-sections show a small amount of spherical porosity that is, overall, lower than the porosity in the density cubes manufactured with the Gaussian beam profile. In contrast to the density cubes manufactured with the Gaussian beam profile, there is no increase in porosity at VEDs > 100 J/mm³. This indicates a broad process window that is more robust against variations of process parameters, thus increasing the process stability. Furthermore, the lack of keyhole porosity at high VEDs affirms that the amount of vaporization is reduced by the donut beam profile. This leads to an increase in maximum feasible energy input.

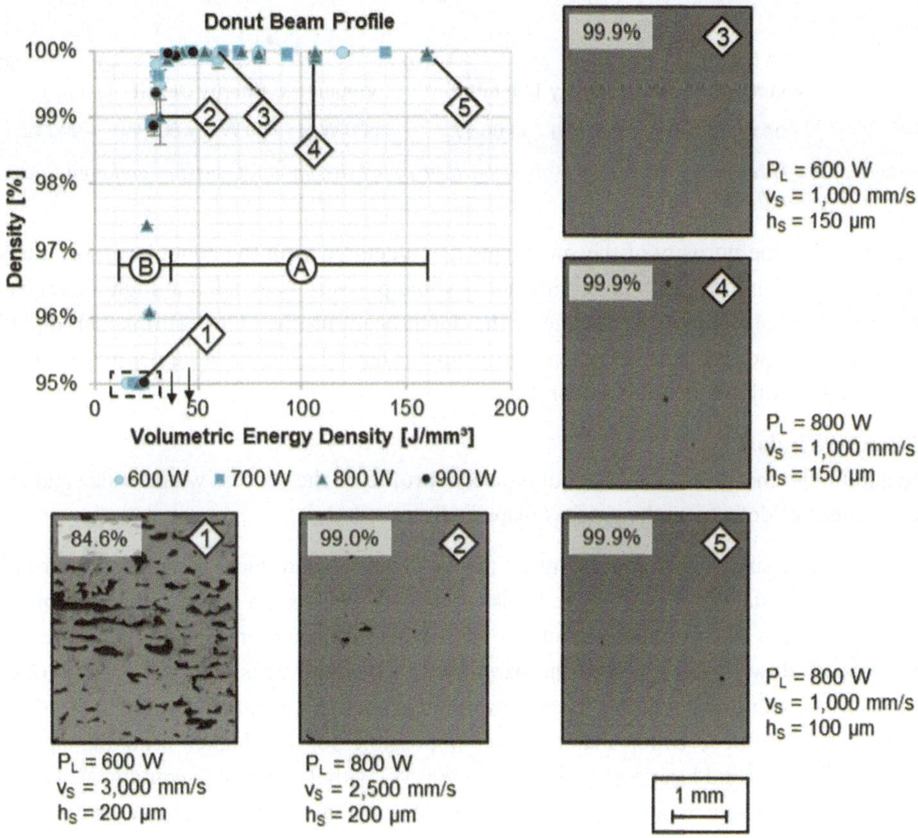

Figure 5-24: Representative micro-sections of specimens manufactured with the donut beam profile in AlSi10Mg.

Furthermore, it is interesting to note that the threshold VED at which lack-of-fusion porosity occurs is very distinct for laser powers between $P_L = 600 - 900$ W and independent of the laser beam profile, scanning speed or hatch distance. Figure 5-25 illustrates that all cubes manufactured with VEDs > 35 J/mm³ have a lack-of-fusion porosity < 1% and that all cubes manufactured with VEDs < 30 J/mm³ have a lack-of-fusion porosity > 1%. This demonstrates a distinct minimum energy input necessary to achieve a sufficient layer bonding and melting of the powder particles for the specific range of laser powers.

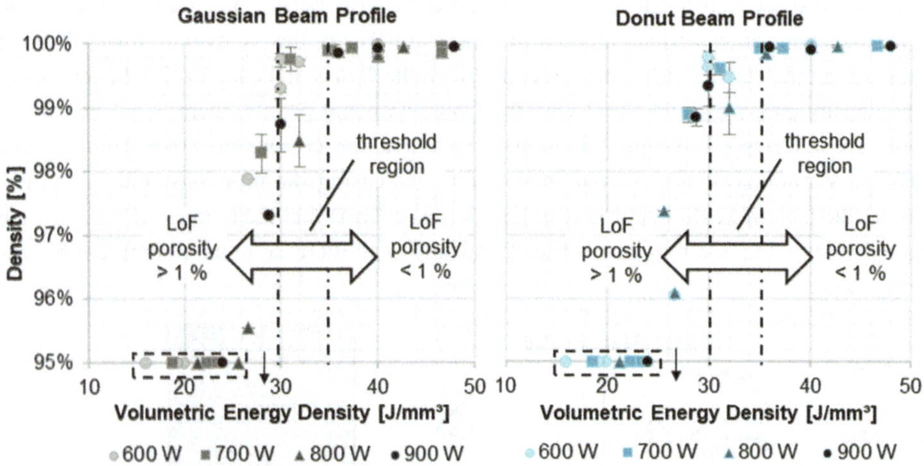

Figure 5-25: Threshold VED for lack-of-fusion (LoF) porosity for the Gaussian and donut beam profile for AlSi10Mg.

The fact that the porosity of the specimens manufactured with the donut beam profile is overall lower and the fact that significant keyhole porosity is avoided at high VEDs can be seen as a considerable advantage of the donut beam profile. This affirms that, due to the reduced vapor pressure, a higher energy input is feasible and that a stabler melt process is attainable with the donut beam profile.

Process Window

In addition to the above, the effect of the beam profile on the process window size and the occurrence of defects on the process maps is discussed.

Figure 5-26 visualizes the process maps for the Gaussian and donut beam profile for different hatch distances, based on the measured densities. One can see that there is a broad process window that produces densities > 99.5% for both beam profiles. Furthermore, the process window visibly exceeds the parameters tested and densities > 99.5% can most probably be achieved at higher laser power and scanning speed than the ones tested for both beam profiles. Unfortunately, it is impossible to test higher scanning speeds (v_{S_MAX} = 3 m/s) and laser powers (P_{L_MAX} = 900 W) with the available equipment. Therefore, it is not feasible to investigate the physical limits of the process experimentally within this thesis. However, a theoretical calculation is done in Chapter 5.3.2.

Gaussian Beam Profile

hs = 100 µm	Scanning Speed [mm/s] 1,000	1,500	2,000	2,500	3,000
Laser Power [W] 300	97.6%	97.8%	95.4%	n/a	n/a
400	99.7%	99.8%	99.5%	n/a	n/a
500	99.9%	100.0%	100.0%	n/a	n/a
600	99.4%	100.0%	100.0%	n/a	n/a
700	97.9%	99.7%	99.9%	n/a	n/a
800	97.8%	99.7%	99.9%	n/a	n/a
900	n/a	n/a	n/a	n/a	n/a

hs = 150 µm	Scanning Speed [mm/s] 1,000	1,500	2,000	2,500	3,000
Laser Power [W] 300	99.3%	96.9%	87.0%	n/a	n/a
400	100.0%	99.9%	98.0%	n/a	n/a
500	99.7%	99.9%	99.9%	n/a	n/a
600	99.8%	99.9%	99.9%	99.7%	97.9%
700	99.3%	99.8%	99.9%	99.9%	99.8%
800	98.9%	99.8%	99.9%	99.9%	99.9%
900	n/a	n/a	n/a	99.9%	99.9%

hs = 200 µm	Scanning Speed [mm/s] 1,000	1,500	2,000	2,500	3,000
Laser Power [W] 300	99.4%	87.3%	65.5%	n/a	n/a
400	99.9%	97.4%	92.5%	n/a	n/a
500	99.8%	99.9%	96.8%	n/a	n/a
600	99.8%	99.9%	99.5%	93.7%	85.3%
700	99.6%	99.8%	99.9%	98.3%	93.4%
800	99.4%	99.8%	99.9%	98.5%	95.5%
900	n/a	n/a	n/a	99.8%	98.7%

hs = 250 µm	Scanning Speed [mm/s] 1,000	1,500	2,000	2,500	3,000
Laser Power [W] 300	n/a	n/a	n/a	n/a	n/a
400	n/a	n/a	n/a	n/a	n/a
500	n/a	n/a	n/a	n/a	n/a
600	n/a	n/a	n/a	83.4%	73.5%
700	n/a	n/a	n/a	85.9%	77.1%
800	n/a	n/a	n/a	94.9%	86.0%
900	n/a	n/a	n/a	97.3%	91.8%

Donut Beam Profile

hs = 100 µm	Scanning Speed [mm/s] 1,000	1,500	2,000	2,500	3,000
Laser Power [W] 300	98.0%	98.0%	93.2%	n/a	n/a
400	99.2%	99.2%	99.0%	n/a	n/a
500	99.6%	99.5%	99.3%	n/a	n/a
600	100.0%	99.9%	99.9%	n/a	n/a
700	100.0%	99.9%	99.9%	n/a	n/a
800	99.9%	100.0%	100.0%	n/a	n/a
900	n/a	n/a	n/a	n/a	n/a

hs = 150 µm	Scanning Speed [mm/s] 1,000	1,500	2,000	2,500	3,000
Laser Power [W] 300	99.3%	97.2%	84.4%	n/a	n/a
400	99.8%	99.8%	98.8%	n/a	n/a
500	99.6%	99.9%	99.7%	n/a	n/a
600	100.0%	100.0%	100.0%	99.5%	96.0%
700	99.9%	100.0%	100.0%	99.9%	99.6%
800	99.9%	100.0%	100.0%	100.0%	99.9%
900	n/a	n/a	n/a	100.0%	99.9%

hs = 200 µm	Scanning Speed [mm/s] 1,000	1,500	2,000	2,500	3,000
Laser Power [W] 300	98.1%	86.5%	73.3%	n/a	n/a
400	99.9%	97.7%	91.7%	n/a	n/a
500	100.0%	99.5%	96.6%	n/a	n/a
600	100.0%	100.0%	99.7%	94.3%	84.6%
700	100.0%	100.0%	99.9%	98.9%	93.4%
800	100.0%	100.0%	100.0%	99.0%	96.1%
900	n/a	n/a	n/a	100.0%	99.4%

hs = 250 µm	Scanning Speed [mm/s] 1,000	1,500	2,000	2,500	3,000
Laser Power [W] 300	n/a	n/a	n/a	n/a	n/a
400				n/a	n/a
500				n/a	n/a
600				86.1%	77.8%
700				89.2%	77.8%
800				97.4%	92.1%
900				98.9%	94.1%

Density
< 98.0 %
< 99.0 %
< 99.5 %
> 99.5 %

Figure 5-26: Process maps for the Gaussian and donut beam profile for different hatch distances hs for AlSi10Mg.

The most relevant difference in the process windows produced by both beam profiles occurs at high VEDs (high laser powers and low scanning speed), where significant keyhole porosity is present for the Gaussian beam profile. At the same time, it is almost non-existent for the donut beam profile (see Figure 5-27). As previously discussed, the higher vapor pressure caused by the Gaussian beam profile is the primary cause of this keyhole porosity [1, 101]. At low VEDs (high scanning speed and hatch distance), lack-of-fusion porosity occurs for both beam profile at similar process parameters at hatch distances of $h_S \geq 150$ µm. At a hatch distance of $h_S \geq 100$ µm, lack-of-fusion porosity only occurred at a laser power of $P_L = 300$ W for the Gaussian beam profile, while a low amount of lack-of-fusion porosity (< 1%) occurred up to a laser power of $P_L = 500$ W for the donut beam profile. This indicates that the process is slightly more energy efficient for the Gaussian beam profile at these parameters. However, these process parameters are linked to low productivity due to the low hatch distance and are not the primary focus of the investigations in this thesis.

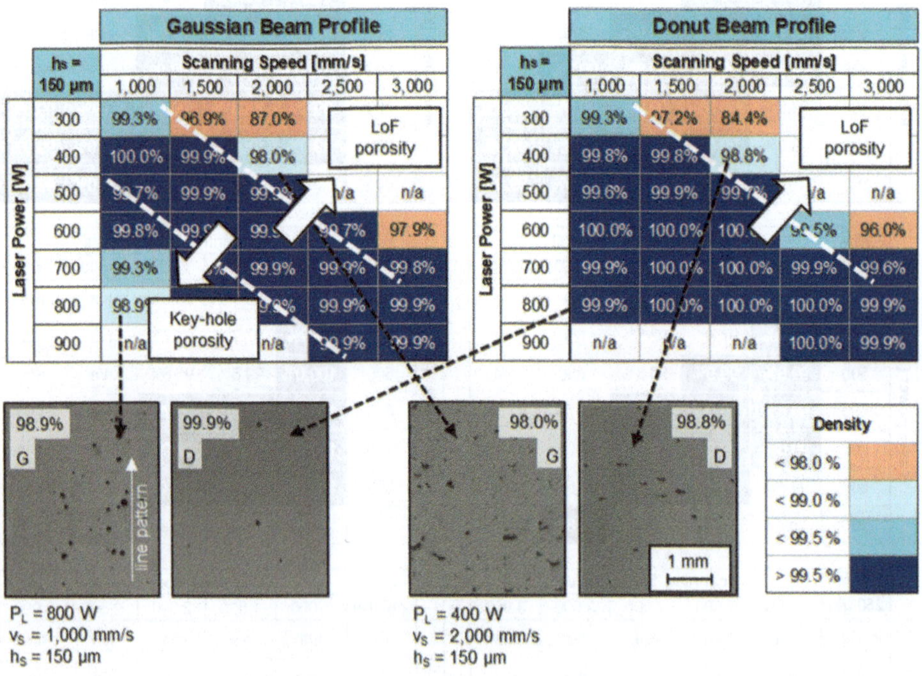

Figure 5-27: Occurrence of lack-of-fusion and keyhole porosity on process maps for a hatch distance $h_S = 150$ µm for AlSi10Mg.

Furthermore, it is interesting to note that within the process window that produces densities ≥ 99.5%, the cubes manufactured with the donut beam profile have higher densities for almost all process parameters. The low amount of spherical porosity present in these specimens can either come from keyhole porosity or hydrogen porosity entrapped in the melt pool. Therefore, this difference can be explained by the lower vapor pressure and lower melt velocities caused by the donut beam profile. The lower vapor pressure reduces the risk of keyhole porosity at the end of scan tracks and the lower melt velocities (see

Chapter 5.1.2) reduce the risk of gas entrapment in the rear section of the melt pool (see Figure 2-23 on page 28) [55, 235, 255].

Spatter Formation

The formation of spatters during the SLM process are a result of vaporization and high melt pool velocities. Spatters are an undesired effect as they harm the process stability and quality of the build [101, 210] e.g., when they interfere with the laser beam [132] or fall on the powder bed at a location where the part is being built [261]. As described in Chapter 2.2.5, there are two different types of spatters: droplet spatters, which are ejected from the melt pool due to overheating and high melt velocities, and entrainment driven spatters, which are formed when surrounding particles are entrained by the metal vapor jet [101, 132, 137].

To investigate the effect of the beam profile on the spatter formation, high-speed videos of the SLM process are recorded. The amount of spatters is analyzed with the help of a MATLAB script that automatically counts the spatters for every single frame, as described in Chapter 4.4 and Appendix A.3. Dark spatters and bright spatters are distinguished by a defined threshold brightness, during the processing of each frame. As particles ejected from the melt pool will have a significantly higher brightness than entrained particles, it is assumed that there is a high correlation between bright spatters and spatters ejected from the melt pool, as well as dark spatters and particles that are entrained due to the metal vapor jet.

Figure 5-28 exemplarily shows binary images of the spatters recorded during the multi-layer SLM process for both beam profiles at the same process parameters. The images demonstrate that the amount of spatters is significantly lower for the donut beam profile than for the Gaussian beam profile. To quantify this difference, Figure 5-29 shows the average amount of spatters for different VED and for different build rates as defined by Equation 5.1 (see page 95). The average amount of spatters per frame varies between 9 - 25 with a mean value of 15.2 for the Gaussian beam profile and between 5 - 16 with a mean value of 9.8 for the donut beam profile. Therefore, the amount of spatters is, on average, more than 35% lower for the donut beam profile. The primary reason for this difference is seen in the fact that the amount of spatters depends to a high degree on the magnitude of vaporization as well as the velocity of the melt flow [31, 132, 137]. The lower vapor pressure caused by the donut beam profile (see Chapter 5.1.3) reduces the amount of entrainment driven spatters, and the lower melt velocities (on average ~20%, see Chapter 5.1.2) reduces the amount of spatters ejected from the melt pool.

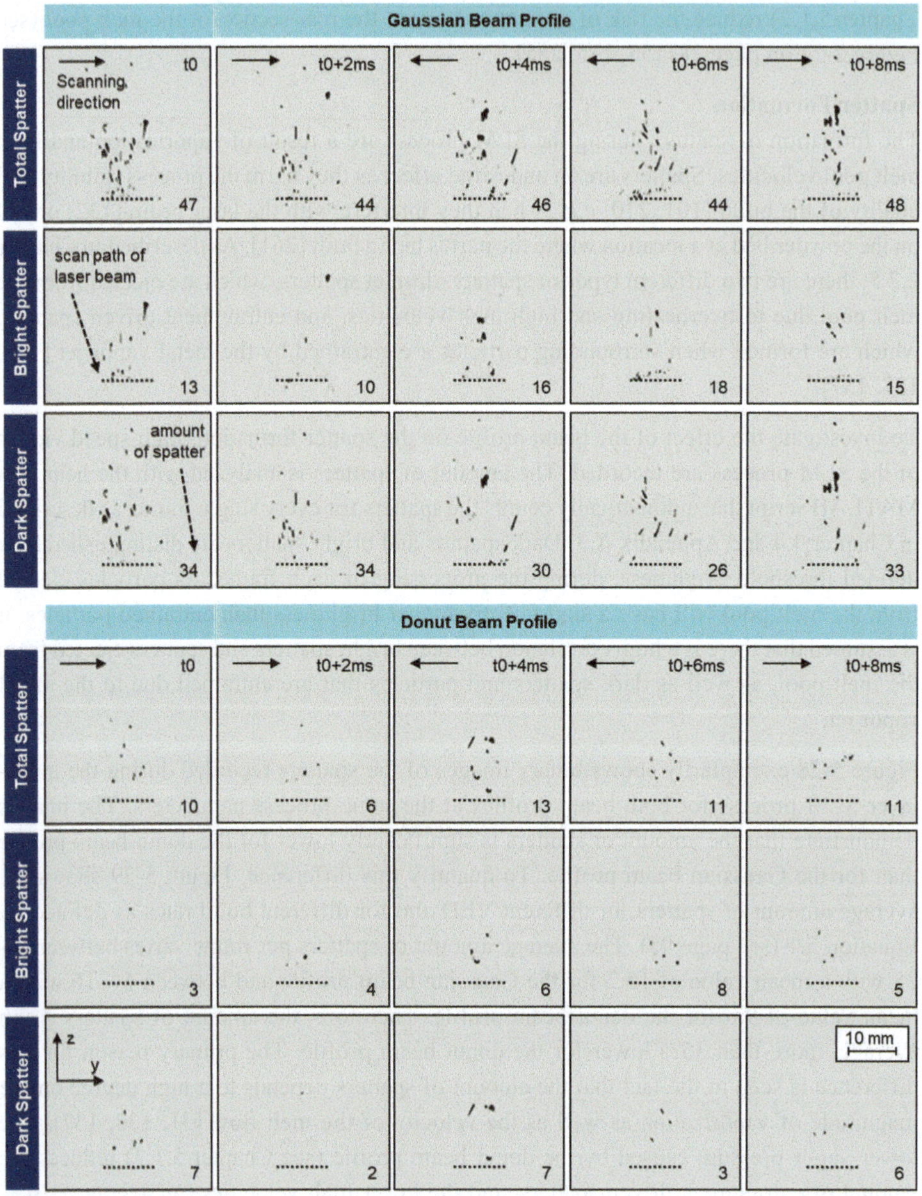

Figure 5-28: Binary images visualizing the spatters occurring during the multi-layer SLM process of AlSi10Mg at a laser power of P_L = 800 W, scanning speed of v_S = 2,500 mm/s and hatch distance of h_S = 200 µm for the Gaussian and donut beam profile. The scanning direction of the laser beam is parallel to the viewing direction as indicated by the dotted line.

Furthermore, it can be noted that the amount of spatters decreases with an increase in VED and increases with an increase in build rate. It might be surprising that the total amount of spatters decreases with an increase in VED at first sight, but this can be explained by the fact that high VEDs are linked to low scanning speeds (see Equation 2.3) and the effect the scanning speed has on spatter formation: First, a low scanning speed reduces the number of particles the laser beam interacts with and potentially can entrain within a certain amount of time. This will consequently lead to a reduction in entrainment-driven spatters (dark spatters). Secondly, low scanning speeds lead to a decrease in melt velocities as discussed in Chapter 5.1.2, thus reducing the particle ejection from the melt pool (bright spatters) [137, 181]. The latter is affirmed by Figure 5-30, which displays an increase in spatters with an increase in scanning speed. The effect of the hatch distance, which influences the VED in the same manner, is not as distinct. As high scanning speeds are linked to high build rates, the discussed effects are also seen as the primary reason for the increase in spatters that occurs with an increase in build rate.

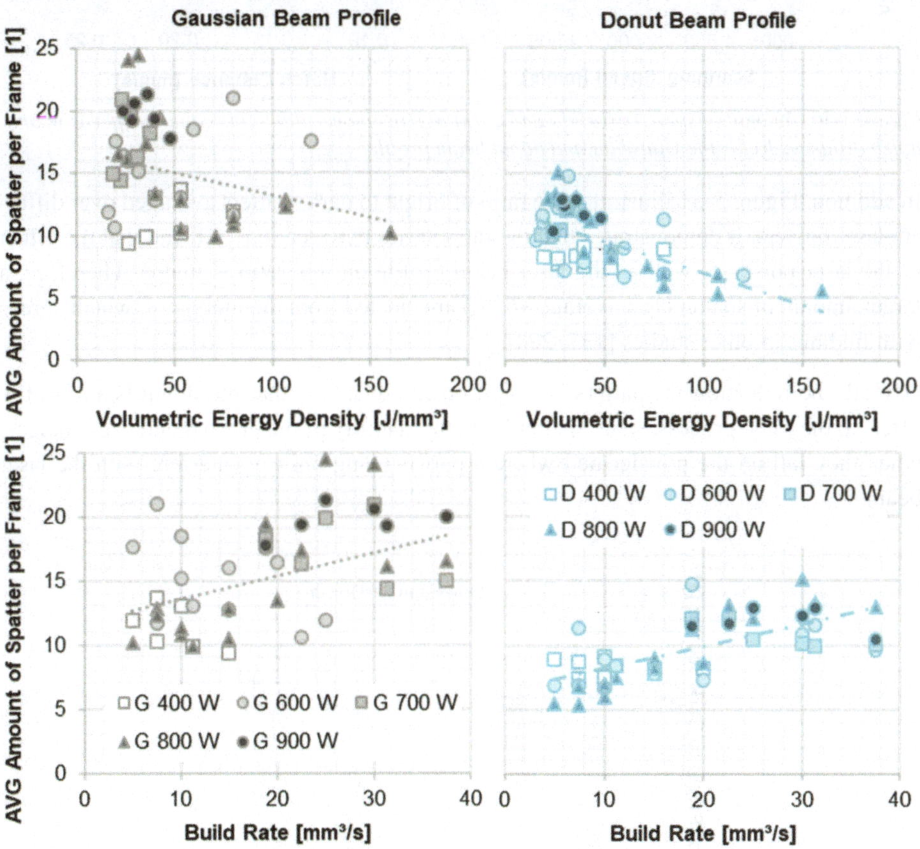

Figure 5-29: Amount of spatters (bright and dark) averaged over 200 frames for different VEDs and build rates for the Gaussian and donut beam profile.

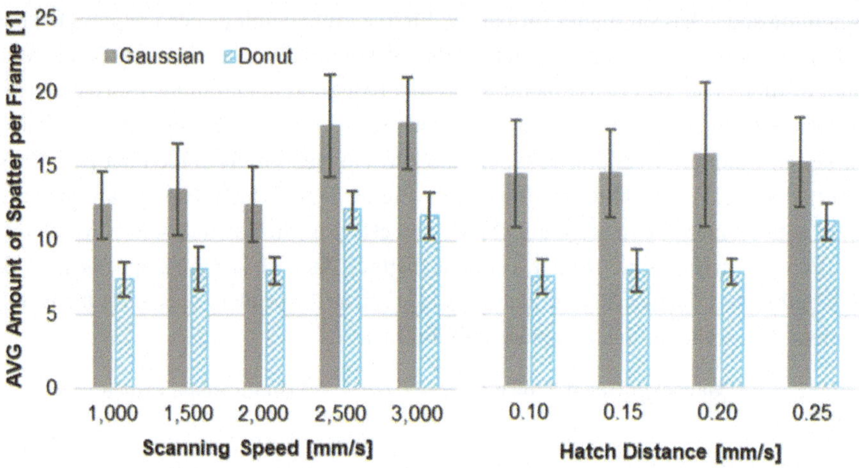

Figure 5-30: Amount of spatters (bright and dark) averaged over different scanning speeds vs and hatch distances hs for the Gaussian and donut beam profile.

In addition, Figure 5-31 illustrates the ratio of bright to dark spatters averaged over different laser powers. It demonstrates that the ratio varies between 0.5 - 0.8 and that the effect of the laser power as well as the laser beam profile on this ratio is minor. Therefore, a larger amount of spatters (on average ~65%) are ejected from the melt pool, which agrees with findings in other studies [137, 224].

Overall, the reduction in spatters is a significant advantage that the donut beam profile offers to the SLM process, since spatter harm the stability of the process and cause defects when they fall on the powder bed where a part is being built or interfere with the laser beam.

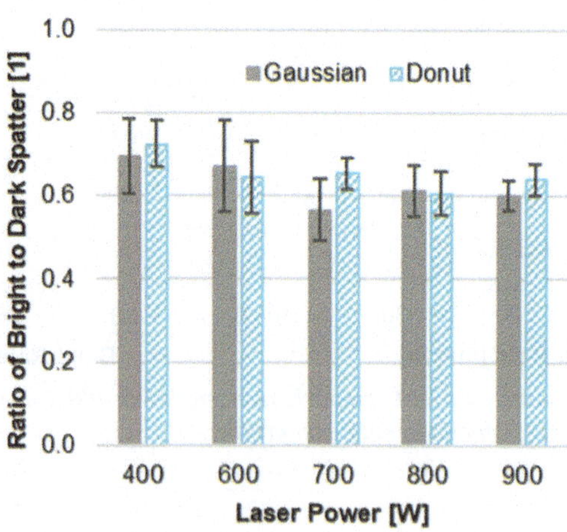

Figure 5-31: Ratio of bright to dark spatters for the Gaussian and donut beam profile, averaged over different laser powers.

5.3.2 Effect on Build Rate

Second to an increase in process stability, an increase in productivity is a significant motivation for the investigations done in this thesis. As the productivity of the process is primarily linked to the build rate, the effect of the laser beam profile on the maximum feasible build rate is investigated in this section.

The process parameters with the highest build rates are linked to high scan speeds, high hatch distances and high layer thicknesses as defined by Equation 5.1 [22]. Figure 5-32 shows the region of highest build rates on the process maps of the previous chapter and Figure 5-33, the specimens density over different build rates.

$$b = v_S \cdot h_S \cdot h_l \tag{5.1}$$

b: Build rate [mm³/s]
v_S: Scanning speed [mm/s]
h_S: Hatch distance [µm]
h_l: Leveling height of build platform [µm]

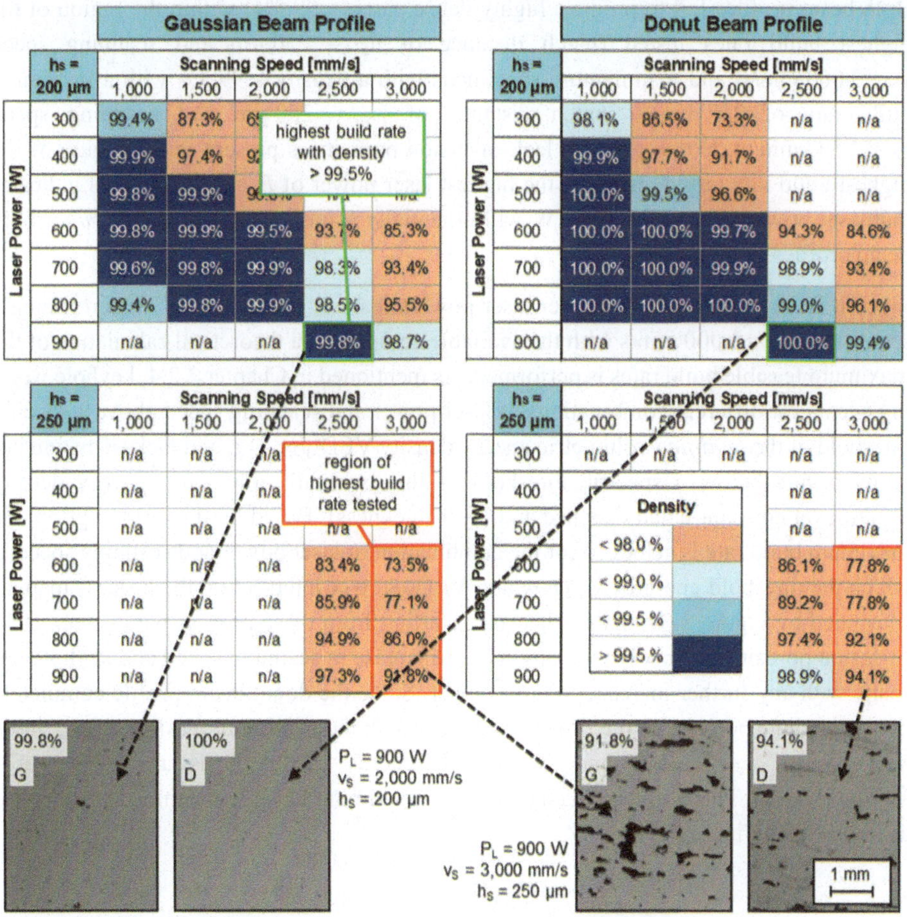

Figure 5-32: Region of highest build rates on process map for AlSi10Mg.

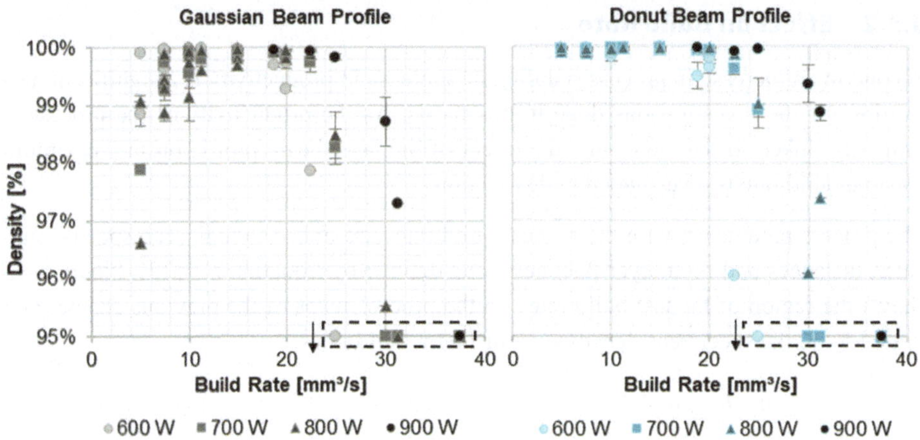

Figure 5-33: Density of AlSi10Mg specimens manufactured with the Gaussian and donut beam profile for different build rates.

Both beam profiles fail to produce highly dense parts (> 99.5%) within the region of the highest build rates tested (hatch distance of h_S = 250 µm and scanning speed v_S = 3.000 mm/s) and manufacture specimens with a density > 99.5% with a maximum build rate of 25 mm³/s (hatch distance of h_S = 200 µm and scanning speed v_S = 2.500 mm/s). However, only lack-of-fusion porosity is present in the region of the highest build rate tested, even for the highest laser power of P_L = 900 W. This indicates that laser powers above P_L = 900 W will potentially achieve densities > 99.5% for both beam profiles.

Since it is impossible to test higher laser powers than P_L = 900 W and higher scanning speeds than v_S = 3,000 mm/s with the available equipment, a theoretical calculation of the maximum feasible build rates is performed. As mentioned in Chapter 2.2.4, keyhole welding occurs for different combinations of laser power and scanning speed above a specific threshold of the *modified* volumetric energy density VED_{MOD_MAX}, and lack-of-fusion porosity occurs below a specific threshold of the "normal" volumetric energy density VED_{MIN} (at constant hatch distance and layer thickness). Based on this presumption, the maximum build rate is achieved for the combination of laser power and scanning speed at which the threshold energies VED_{MIN} and VED_{MOD_MAX} intersect on the process map, as illustrated by Figure 5-34. This is because, starting from these process parameters, lack-of-fusion porosity will occur with any further increase in scanning speed and keyhole porosity with any further increase in laser power. Since the donut beam profile completely avoids keyhole porosity in the process regime tested, the maximum feasible energy input $VED_{MOD\_MAX\_D}$ is higher than the feasible energy of the Gaussian beam profile $VED_{MOD\_MAX\_G}$. This will lead to an increase in maximum feasible scanning speed v_{S_MAX} as indicated in Figure 5-34.

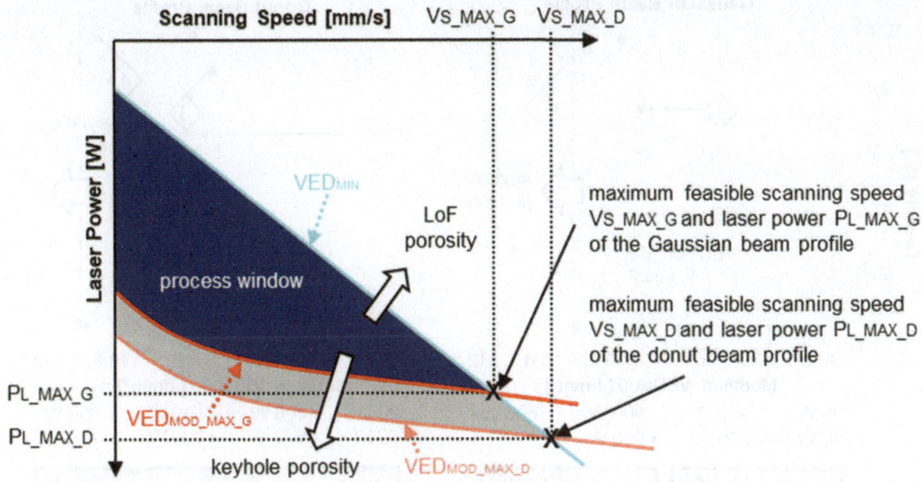

Figure 5-34: Theoretical process window with minimum volumetric energy density VED_{MIN} and maximum modified volumetric energy density VED_{MOD_MAX} for both beam profiles.

Based on the assumptions above, the maximum feasible energy input and scanning speed is calculated for a hatch distance of h_S = 200 µm and layer thickness of h_l = 50 µm for both beam profiles. To estimate the threshold energy VED_{MOD_MAX} above which keyhole welding occurs for both beam profiles, Figure 5-35 shows the specimens density for different VED_{MOD} as defined by Equation 2.4 (page 16). The figure shows that keyhole welding occurs above values of VED_{MOD} = 9 J/mm^3 for the Gaussian beam profile and that the threshold energy $VED_{MOD\_MAX\_D}$ is not reached with the process parameters tested for the donut beam profile. Therefore, a value of $VED_{MOD\_MAX\_D}$ = 11 J/mm^3 is used for the calculations. However, it has to be kept in mind that the actual value for $VED_{MOD\_MAX\_D}$ might be even higher, which would lead to higher build rates for the donut beam profile. The threshold VED_{MIN} for lack-of-fusion porosity is displayed in Figure 5-25 (page 88) and is 35 J/mm^3 for both beam profiles.

Based on the VED_{MIN} and VED_{MOD_MAX}, the maximum scanning speed v_{S_MAX} is calculated, as shown in Equation 5.5, by substituting the laser power P_L (Equation 5.3) in Equation 5.4.

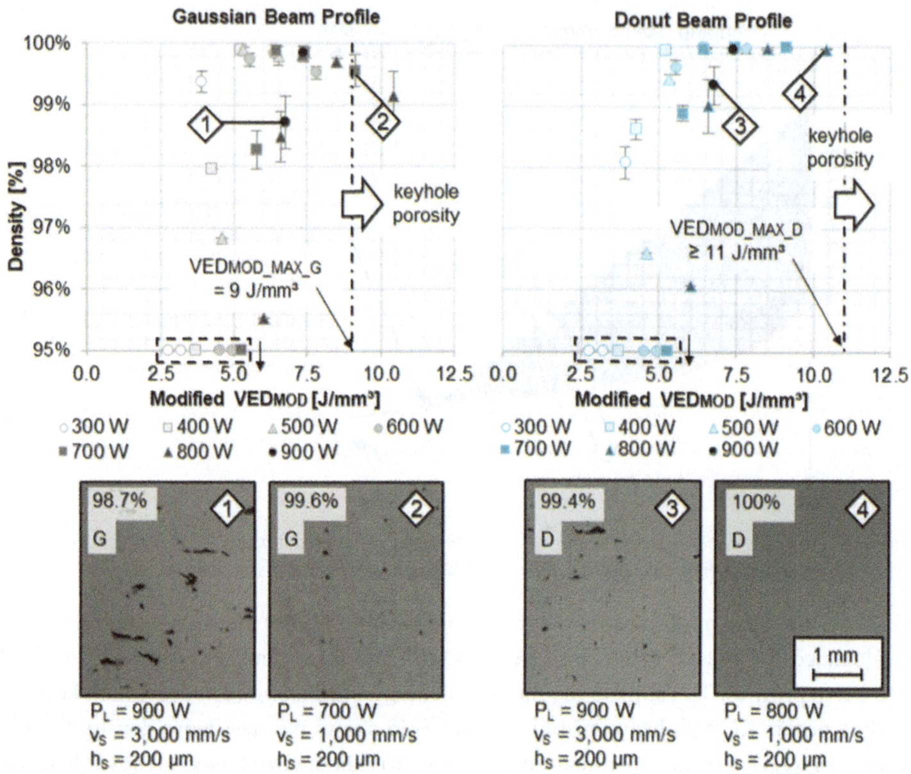

Figure 5-35: Density of specimens manufactured in AlSi10Mg with the Gaussian (G) and donut (D) beam profile and hatch distance of hs = 200 µm for different VEDMOD with the threshold value for VEDMOD_MAX at which keyhole welding occurs.

$$VED = \frac{P_L}{v_S \cdot h_S \cdot h_l} \tag{5.2}$$

$$\rightarrow P_L = v_S \cdot h_S \cdot h_l \cdot VED \tag{5.3}$$

$$VED_{MOD} = \frac{a_{fres}}{h_S}\sqrt{\frac{C_p \cdot \rho}{4 \cdot k \cdot d_f}} \cdot \frac{P_L}{\sqrt{v_S}} = a_{fres} \cdot h_l \sqrt{\frac{C_p \cdot \rho \cdot v_S}{4 \cdot k \cdot d_f}} \cdot VED \tag{5.4}$$

$$\rightarrow v_{S\_MAX} = \frac{4 \cdot k \cdot d_f}{C_p \cdot \rho \cdot v_S \cdot a_{fres}^2 \cdot h_l^2}\left(\frac{VED_{MOD\_MAX}}{VED_{MIN}}\right)^2 \tag{5.5}$$

VED:	Volumetric energy density [J/mm³]	VED_{MOD}:	Modified VED [J/mm³]
VED_{MOD_MAX}:	Maximum VEDMOD [J/mm³]	VED_{MIN}:	Minimum VED [J/mm³]
P_L:	Laser power [W]	v_S:	Scanning speed [mm/s]
v_{S_MAX}:	Maximum scanning speed [mm/s]	h_S:	Hatch distance [µm]
h_l:	Layer thickness [µm]	k:	Thermal conductivity [W/(m·K)]
d_f:	Focus diamter of laser beam [µm]	C_p:	Specific heat capacity [J/kgK]
ρ:	Density of material [kg/m³]	a_{fres}:	Absorption coefficient [1]

Based on Equations 5.3 and 5.5, and assuming constant material properties of k = 100 W/(m·K), C_P = 1,030 J/kgK, ρ = 2,400 kg/m³ and an absorption coefficient of a_{fres} = 40% (see Table 3-1 on page 47), the maximum feasible scanning speed v_{S_MAX}, build rate b_{MAX} and laser power P_{L_MAX} for the Gaussian and donut beam profiles are calculated (see equations 5.6 - 5.11). The results reveal that the maximum feasible build rate is b_{MAX_G} = 37 mm³/s for the Gaussian beam profile and b_{MAX_D} = 59 mm³/s for the donut beam profile and is, therefore, significantly higher than the build rate of 25 mm³/s demonstrated in the experiments. Based on this, it can be estimated that the donut beam profile offers the potential to increase the build rate by 60%.

$$v_{S\_MAX\_G} = \frac{4 \cdot k \cdot d_f}{C_p \cdot \rho \cdot a_{fres}^2 \cdot h_l^2} \left(\frac{VED_{MOD\_MAX\_G}}{VED_{MIN\_G}}\right)^2 = 3{,}744\ \frac{mm}{s} \tag{5.6}$$

$$v_{S\_MAX\_D} = \frac{4 \cdot k \cdot d_f}{C_p \cdot \rho \cdot a_{fres}^2 \cdot h_l^2} \left(\frac{VED_{MOD\_MAX\_G}}{VED_{MIN\_G}}\right)^2 = 5{,}994\ \frac{mm}{s} \tag{5.7}$$

$$b_{MAX\_G} = v_{S\_MAX\_G} \cdot h_S \cdot h_l = 37\ \frac{mm^3}{s} = 133\ \frac{cm^3}{h} \tag{5.8}$$

$$b_{MAX\_D} = v_{S\_MAX\_D} \cdot h_S \cdot h_l = 59\ \frac{mm^3}{s} = 212\ \frac{cm^3}{h} \tag{5.9}$$

$$P_{L\_MAX\_G} = v_{S\_MAX\_G} \cdot h_S \cdot h_l \cdot VED_{MIN} = 1{,}310\ W \tag{5.10}$$

$$P_{L\_MAX\_D} = v_{S\_MAX\_D} \cdot h_S \cdot h_l \cdot VED_{MIN} = 2{,}079\ W \tag{5.11}$$

$v_{S\_MAX\_G}$:	Maximum feasible scanning speed for Gaussian beam profile for AlSi10Mg [mm/s]		
$v_{S\_MAX\_D}$:	Maximum feasible scanning speed for donut beam profile for AlSi10Mg [mm/s]		
b_{MAX_G}:	Maximum feasible build rate for Gaussian beam profile for AlSi10Mg [mm³/s]		
b_{MAX_D}:	Maximum feasible build rate for donut beam profile for AlSi10Mg [mm³/s]		
$P_{L\_MAX\_G}$:	Maximum feasible laser power for Gaussian beam profile for AlSi10Mg [W]		
$P_{L\_MAX\_D}$:	Maximum feasible laser power for donut beam profile for AlSi10Mg [W]		
$VED_{MOD\_MAX\_G}$:	Maximum *modified* VED_{MOD} for Gaussian beam profile for AlSi10Mg (9 J/mm³)		
$VED_{MOD\_MAX\_G}$:	Maximum *modified* VED_{MOD} for donut beam profile for AlSi10Mg (11 J/mm³)		
VED_{MIN}:	Minimum VED for Gaussian and donut beam profile for AlSi10Mg (35 J/mm³)		
k:	Thermal conductivity (100 W/(m·K))	d_f:	Focus diameter of laser (140 µm)
C_p:	Specific heat capacity (1,030 J/kgK)	ρ:	Density of material (2,400 kg/m³)
a_{fres}:	Absorption coefficient (0.4)	h_l:	Layer thickness (50 µm)
h_S:	Hatch distance (200 µm)		

It must be noted that the increase in potential energy input by the donut beam profile of 60% (see Equations 5.10 and 5.11) is more significant than indicated by the numerical results presented in Chapter 5.1. It is assumed that the main reason for this is the fact that several important phenomena that are relevant for the defect formation in the multi-layer SLM process are not considered in the simulation of the single tracks, such as hatch-distance, pre-heating due to neighboring tracks, denudation, or hydrogen solubility. Therefore, it is expected that the calculations in this chapter that are based on the experimental results, are more valid.

The circumstance that the maximum scanning speed and build rate can be increased by 60% with the donut beam profile can be seen as a significant advantage that leads to relevant manufacturing cost and time reductions. However, the effect on the overall productivity is lower than 60% due to the time necessary to "jump" between single tracks and the recoating time of each layer. Therefore, the effect on the manufacturing time and costs is calculated in Chapter 6.1 for selected use cases.

5.3.3 Comparison to Single Track Results

Comparing the results of the multi-layer SLM process to the results of the single-track experiments reveals some important considerations and limitations that have to be taken into account when analyzing the single track results. First, a direct correlation between the quality of the single tracks (top view analysis, see Figure 5-16 on page 78) and the ability to manufacture fully dense parts (see Figure 5-26 on page 89) is not possible [239]. This is because several parameters and phenomena that are critical for the multi-layer SLM process are not considered in the single-track experiments, such as hatch-distance, denudation, and pre-heating due to neighboring tracks. Second, the ability to manufacture symmetric cross-sections seems to have more validity in predicting the possibility to manufacture dense parts than the appearance of the single tracks (compare Figure 5-16 and Figure 5-26).

5.3.4 Summary of Results

To investigate the effect of the beam profile on the multi-layer SLM process for AlSi10Mg, density cubes were manufactured and the porosity was analyzed by microsections. Furthermore, high-speed videos of the process were recorded to study the spatter formation. Based on the results, the effect of the beam profile on the process stability and build rate was discussed. It can be summarized that there is a considerable effect of the beam profile on the multi-layer SLM process of AlSi10Mg and that the donut beam profile offers several advantages over the Gaussian beam profile.

Although both beam profiles can manufacture highly dense parts (> 99.5%) with a wide range of process parameters, the density of the specimens manufactured with the donut beam profile is higher for almost all parameters. At low VEDs (< 35 J/mm^3), lack-of-fusion porosity occurs at similar parameters for both beam profiles, while at high VEDs (> 100 J/mm^3), spherical porosity only occurs for the Gaussian beam profile. This confirms that higher energy inputs and larger process windows are possible with an optimized beam profile, such as the donut beam profile. Furthermore, the amount of spatters caused by the donut beam profile is, on average, 35% lower than the amount caused by the Gaussian beam profile. This demonstrates a stabler melting process and lower risk of defects due to spatters that fall on the powder bed or interfere with the laser beam. The primary reason for the differences is seen in the reduced vapor pressure and lower melt velocities caused by the donut beam profile, leading to a stabler melting process.

In addition, the beam profile has a considerable effect on the productivity of the process. Since it was not possible to test the physical limits of the process with the available equipment, calculations on the maximum feasible build rates were done, based on the experimental results. The results reveal that an increase in build rate by 60% is possible with the donut beam profile, compared to the Gaussian beam profile. However, experimental investigations are necessary in the future to underpin these results.

5.4 Effect on Multi-Layer SLM process of AlSi16Sc

To explore if the effect of the beam profile depends on the alloy composition, additional multi-layer SLM experiments are conducted with a high strength AlSi16Sc aluminum alloy. The alloy is more challenging to process in SLM than AlSi10Mg, particularly due to a reduction in hydrogen solubility.
For the investigations, the effect of the beam profile on the process stability (Chapter 5.4.1) and build rate (Chapter 5.4.2) is discussed and compared to the results obtained for AlSi10Mg. Furthermore, a summary of the results is provided (Chapter 5.4.3).

It must be mentioned that the investigated process window is smaller, compared to AlSi10Mg, due to the limited amount of AlSi16Sc powder that was available.

5.4.1 Effect on Process Stability

First, the laser beam profile's effect on the feasible energy input is discussed and afterward on the process window. Both are crucial parameters describing the process stability of the SLM process.

Feasible Energy Input
Figure 5-36 displays the density of the AlSi16Sc specimens depending on different VEDs. The corresponding micro-sections are shown in Appendix A.8 for reference. It can be seen that the density of the cubes manufactured with the donut beam profile is higher for almost all process parameters and that the amount of cubes with densities > 99.5% is significantly higher compared to the ones manufactured by the Gaussian beam profile (43 compared to 25). With the exception of the specimens manufactured with a laser power of $P_L = 300$ W, dense parts (density > 99%) are consistently manufactured with VEDs between 35 - 65 J/mm^3 (region A) for the Gaussian beam profile, and significantly porosity is found below and above these threshold VEDs (region B and C). For the donut beam profile, significant porosity is only found for VEDs below 35 J/mm^3 (region B). Therefore, the results are comparable to the ones obtained for AlSi10Mg, with the difference that the region A that produces dense parts is smaller for the Gaussian beam profile for AlSi16Sc (see Figure 5-21 on page 83).

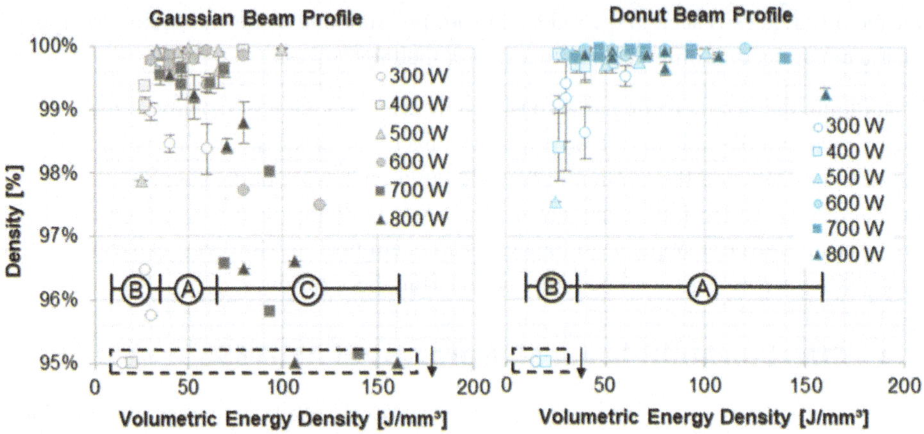

Figure 5-36: Density of specimens manufactured in AlSi16Sc for the Gaussian and donut beam profile for different VEDs.

Figure 5-37 exemplarily shows selected micro-sections of the specimens manufactured with the Gaussian beam profile and Figure 5-38 with the donut beam profile. As for AlSi10Mg, lack-of-fusion porosity occurs at VEDs < 35 J/mm^3 (region B) for both beam profiles and a small amount of spherical porosity is present in region A. Within the region A, line patterns can partly be identified for the specimens manufactured with the Gaussian beam profile. This suggests that the porosity in this region comes at least partially from collapsing keyholes at the end of scan tracks. For VEDs > 65 J/mm^3, a significant increase in spherical porosity can be found for the specimens manufactured with the Gaussian beam profile, which is considerably higher than for AlSi10Mg. For the donut beam profile, only the specimen manufactured with the highest VED of 160 J/mm^3 shows a small amount of spherical porosity.

To understand whether the increase in porosity, compared to AlSi10Mg, comes from key-hole porosity or gas-porosity, a more detailed analysis is necessary. Since AlSi16Sc is a newly developed material, no exact data are available for the relevant material properties that influence the defect generation, such as hydrogen solubility, viscosity and surface tension gradient. Nevertheless, a qualitative analysis is done, based on the effect of the main alloying elements of AlSi10Mg and AlSi16Sc (silicon, magnesium and scandium) on aluminum[4]:

- silicon substantially decreases the hydrogen solubility as well as the viscosity of aluminum and increases the surface tension gradient [12, 46, 85],
- magnesium increases the hydrogen solubility as well as the viscosity [12, 46, 85],
- scandium increases the viscosity [23].

[4] The influence of the alloying elements on other material properties such as density and conductivity are ignored in this context as they are expected to have a minor influence.

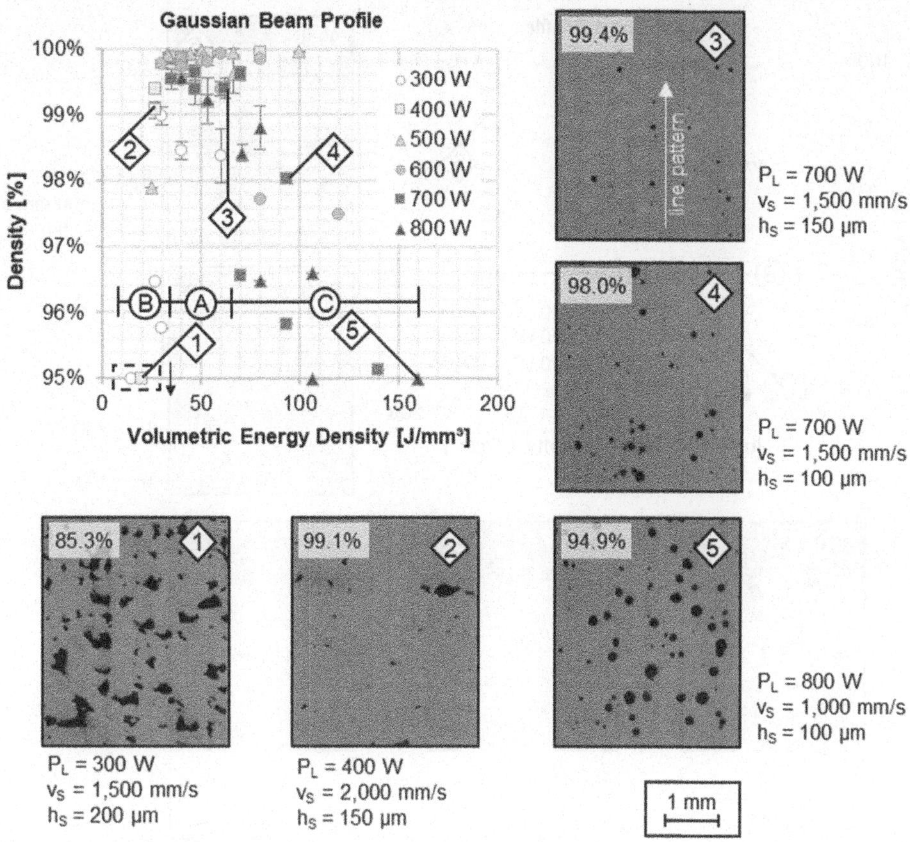

Figure 5-37: Representative micro-sections of specimens manufactured with the Gaussian beam profile in AlSi16Sc.

Based on the described effects as well as the relative change in the constitution of the alloying elements (see Table 5-3), it is expected that the hydrogen solubility and viscosity are considerably lower and that the surface tension gradient is higher for AlSi16Sc, compared to AlSi10Mg. The fact that the hydrogen solubility is lower suggests that the increase in porosity comes from an increase in hydrogen (gas) porosity. The amount of hydrogen in the melt pool will mainly come from hydrogen on the powder particle surface. Therefore, the amount of hydrogen that is potentially entrapped in the melt pool is similar for both materials [147, 235]. Due to the increased hydrogen solubility in the liquid phase, which is about 50 times higher than the solubility in solid aluminum, most of the hydrogen will dissolve in the melt pool and precipitate during solidification [55, 235, 258]. However, the nucleation and growth of hydrogen gas bubbles in the melt pool will be similar for both beam profiles, when one neglects other phenomena. Since no relevant increase in porosity is observed in the specimens manufactured with the donut beam profile (see Figure 5-38), the increase in porosity cannot solely be explained by the lower hydrogen solubility of AlSi16Sc.

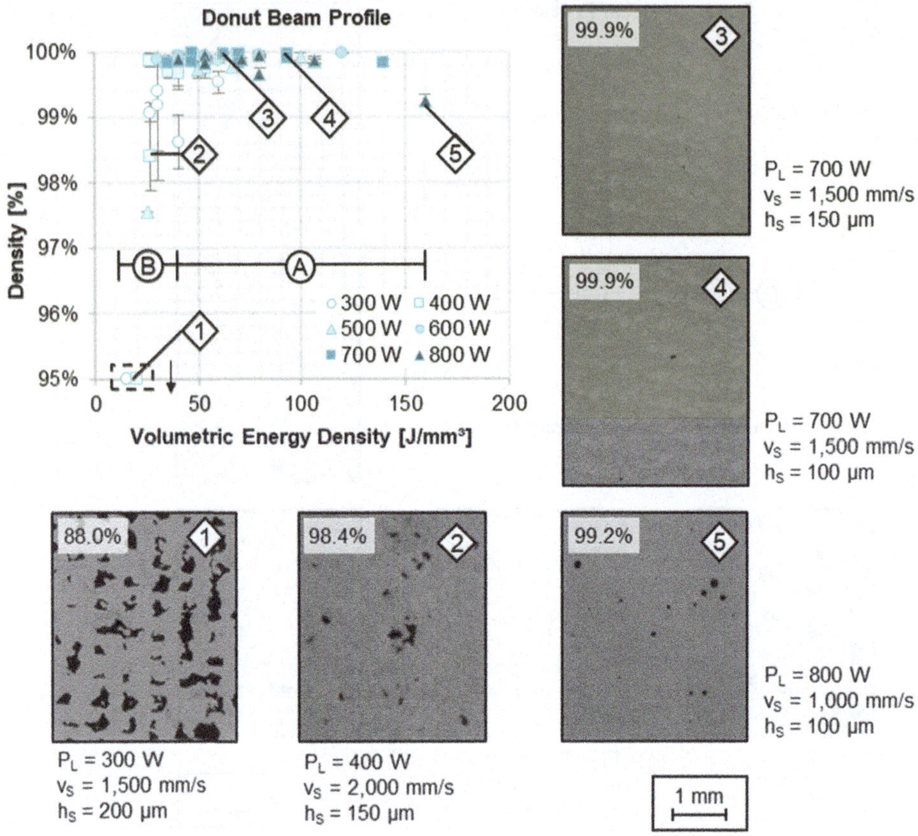

Figure 5-38: Representative micro-sections of specimens manufactured with the donut beam profile in AlSi16Sc.

Table 5-3: Nominal chemical composition of main alloying elements of AlSi10Mg and AlSi16Sc and their effect on the viscosity, surface tension gradient and hydrogen solubility.

n/a : not applicable x : no data or unclear behavior + : increase - : decrease		Aluminum	Silicon	Magnesium	Scandium
Nominal concentration	**AlSi10Mg**	bal.	10%	0.33%	0%
	AlSi16Sc	bal.	16%	0%	0.4%
	Viscosity	n/a	-	+	+
	Surface tension gradient $\Delta\sigma/\Delta T$	n/a	+	x	x
	Hydrogen solubility	n/a	-	+	x
	Melting point [K]	933	1,687	923	1,814
	Boiling point [K]	2,470	3,538	1,364	3,109

Consequently, the cause of the increased porosity has to be found in the combination of the decreased hydrogen solubility and the difference in the melt pool dynamics. Simulations are performed to understand the effect of the modified viscosity and surface tension gradient on the melt pool. Therefore, the viscosity is lowered by 10% and the surface tension gradient is increased by 60% compared to AlSi10Mg (based on an interpolation of the surface tension data of AlSi10Mg and AlSi20 [47])[5]. Figure 5-39 compares the melt pool characteristics for AlSi10Mg and AlSi16Sc exemplarily for a laser power of P_L = 700 W and different scanning speeds, based on the numerical results. In addition, Figure 5-40 visualizes the corresponding surface melt flows. It is shown that the lower viscosity and higher surface tension gradient of AlSi16Sc lead to a small increase in melt pool width, higher melt velocity and slightly lower vapor pressure. The higher melt velocity results from the higher surface tension gradient that strengthens the thermo-capillary flow and the lower viscosity that attenuates the drag of the fluid [74]. Moreover, the higher melt pool width can be explained by the strengthened thermo-capillary flow [97, 254]. Furthermore, the amplified melt flow increases the convective heat transfer away from the laser interaction zone and reduces the peak temperatures. Consequently, this leads to a reduction of vapor pressure. In addition, the lower viscosity reduces the resistance of the fluid against the vapor pressure and increases the instabilities in the melt pool [76, 236]. This explains why the melt pool depth is not affected significantly, despite the lower vapor pressure.

Based on the above, it is assumed that the increase in porosity comes from both an increase in keyhole as well as gas-porosity. Firstly, the lower viscosity of the melt, compared to AlSi10Mg, strengthens the melt pool instabilities and, therefore, the number of gas bubbles due to keyhole collapsing. This type of defect will predominantly occur for the Gaussian beam profile due to the higher vapor pressure. Secondly, the lower hydrogen solubility increases the nucleation and growth of hydrogen gas bubbles in the vicinity of the solidification front (see Figure 2-23 on page 28). Due to the higher temperatures caused by the Gaussian beam profile and the fact that the hydrogen solubility increases with temperature, more hydrogen is dissolved in the melt pool caused by the Gaussian beam profile. Therefore, the nucleation of hydrogen gas bubbles is slightly stronger for the Gaussian beam profile. Although the lower viscosity of the fluid theoretically eases the escape of the gas bubbles [236], they are expected to be entrapped in the melt pool caused by the Gaussian beam profile due to the higher melt velocities [20, 37, 244]. For the donut beam profile, it is assumed that the hydrogen gas bubbles can escape the melt pool due to the less significant and more homogeneous melt flow. Lastly, it has to be mentioned that the generation of gas porosity due to evaporation of low boiling constitutes is not seen as a major cause for the increase in porosity, since the boiling temperatures of the main alloying elements of AlSi16Sc are higher than the boiling temperature of aluminum (see Table 5-3).

[5] all other material properties were assumed to be equivalent to AlSi10Mg due to missing data and/or minor influence on the melt pool dynamics.

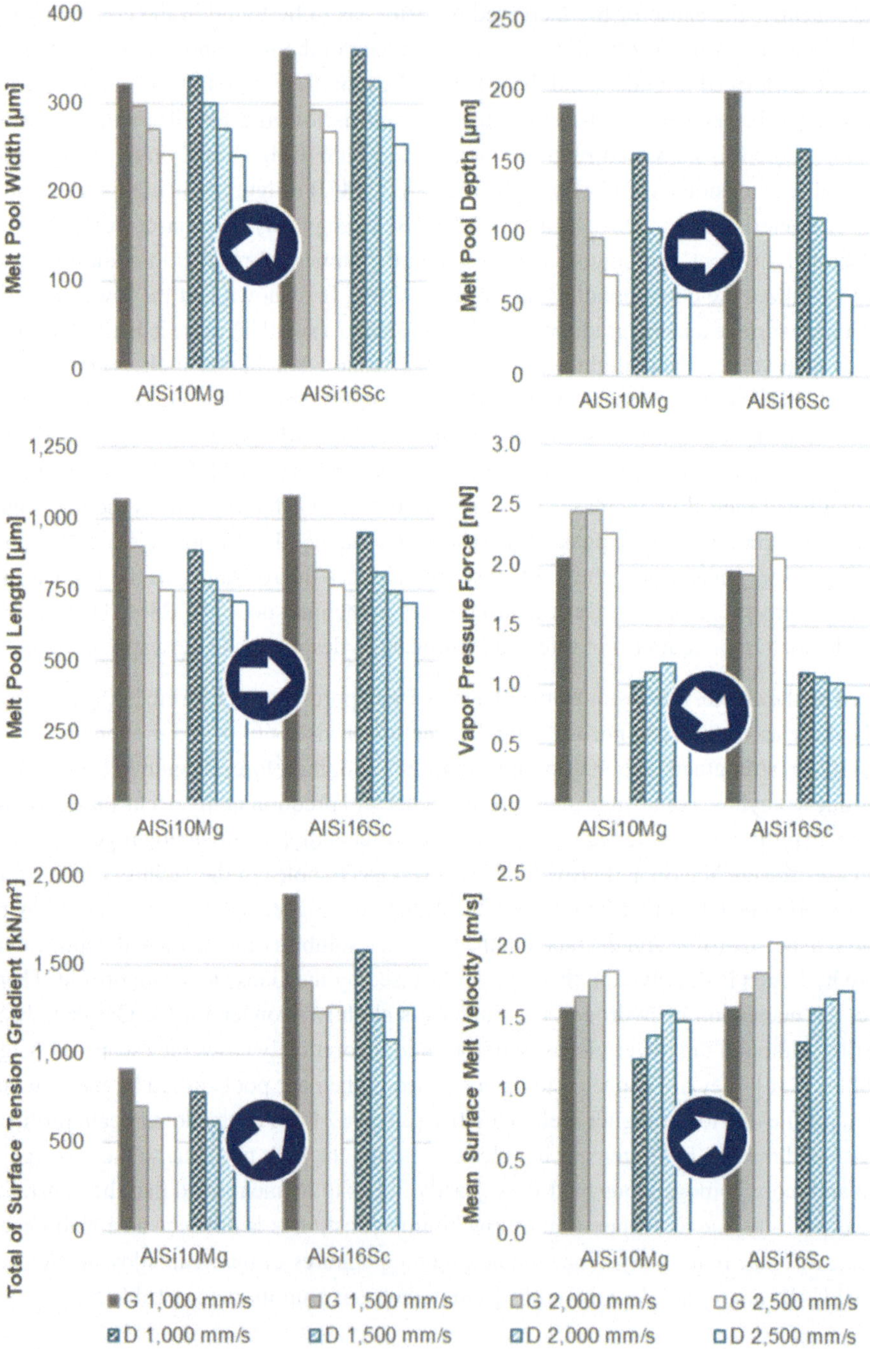

Figure 5-39: Difference in melt pool characteristics for a laser power of P_L=700 W and different scanning speeds for AlSi10Mg and AlSi16Sc.

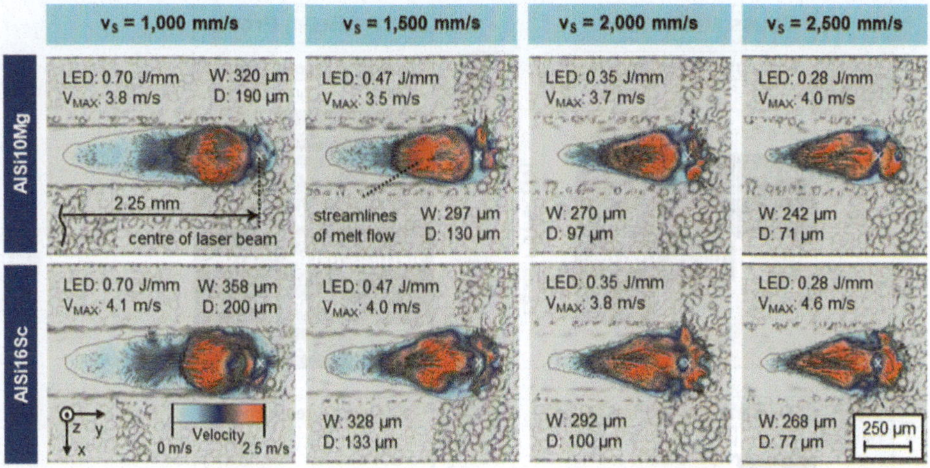

Figure 5-40: Comparison of melt pool surface velocities (color) and flow patterns (streamlines) caused by the Gaussian beam profile for a laser power of P_L=700 W and different scanning speeds at position y=2.25 mm in scanning direction for AlSi10Mg and AlSi16Sc (W: Width; D: Depth).

Process Window

In addition to the above, the effect of the beam profile on the process window size and the occurrence of defects on the process maps is discussed.

Figure 5-41 visualizes the process windows for the Gaussian and donut beam profile for different hatch distances, based on the measured densities. The process window that produces densities > 99.5% is significantly larger for the donut beam profile, compared to the Gaussian beam profile, and exceeds the limits of the parameters tested. As for AlSi10Mg, it is reasonable to assume that dense parts can be manufactured with higher scanning speeds and laser powers than the ones tested. In addition, Figure 5-42 shows that lack-of-fusion porosity is detected for low VEDs (low laser powers and high scanning speed) for both beam profiles. In contrast, significant keyhole porosity is only observed for the Gaussian beam profile at medium to high VEDs. Interestingly, the process window that produces densities > 99.5% is similar, compared to AlSi10Mg, for the donut beam profile, while it is considerably smaller for the Gaussian beam profile.

As explained before, the primary reasons for the higher amount of porosity in the specimens manufactured with the Gaussian beam profile are seen in the increased vapor pressure, the stronger melt flow, and the higher amount of dissolved hydrogen in the melt pool due to higher melt pool temperatures. The larger process window of the donut beam profile that produces densities > 99.5% is a clear advantage and demonstrates a stabler melting process.

Gaussian Beam Pr.

Laser Power [W]	hs = 100 µm	Scanning Speed [mm/s] 1,000	1,500	2,000
	300	98.4%	98.5%	95.8%
	400	99.9%	99.9%	99.7%
	500	100.0%	99.9%	100.0%
	600	97.5%	99.9%	99.9%
	700	95.1%	98.0%	99.6%
	800	94.4%	96.6%	98.8%

Laser Power [W]	hs = 150 µm	Scanning Speed [mm/s] 1,000	1,500	2,000
	300	99.5%	96.5%	86.4%
	400	99.9%	99.9%	99.1%
	500	99.6%	99.9%	99.9%
	600	97.7%	99.8%	99.9%
	700	95.8%	99.4%	99.6%
	800	94.9%	98.4%	99.2%

Laser Power [W]	hs = 200 µm	Scanning Speed [mm/s] 1,000	1,500	2,000
	300	99.0%	88.2%	70.9%
	400	99.9%	99.4%	91.4%
	500	99.8%	99.9%	97.9%
	600	99.4%	99.7%	99.8%
	700	96.6%	99.4%	99.5%
	800	96.5%	99.2%	99.6%

Donut Beam Profile

Laser Power [W]	hs = 100 µm	Scanning Speed [mm/s] 1,000	1,500	2,000
	300	99.5%	99.7%	99.2%
	400	99.9%	99.9%	99.9%
	500	99.9%	100.0%	100.0%
	600	100.0%	100.0%	99.9%
	700	99.8%	100.0%	100.0%
	800	99.3%	99.9%	99.9%

Laser Power [W]	hs = 150 µm	Scanning Speed [mm/s] 1,000	1,500	2,000
	300	98.6%	99.1%	93.6%
	400	99.7%	99.7%	98.4%
	500	99.8%	99.9%	99.9%
	600	99.9%	99.9%	100.0%
	700	99.9%	100.0%	100.0%
	800	99.9%	99.9%	99.9%

Laser Power [W]	hs = 200 µm	Scanning Speed [mm/s] 1,000	1,500	2,000
	300	99.4%	88.7%	82.9%
	400	99.7%	99.9%	94.9%
	500	99.7%	99.9%	97.5%
	600	99.9%	99.9%	99.9%
	700	99.8%	99.8%	99.8%
	800	99.7%	99.8%	99.9%

region of highest build rate tested and density > 99.5 %

Density
< 98.0 %
< 99.0 %
< 99.5 %
> 99.5 %

Figure 5-41: Process maps for the Gaussian and donut beam profile for different hatch distances hs for AlSi16Sc.

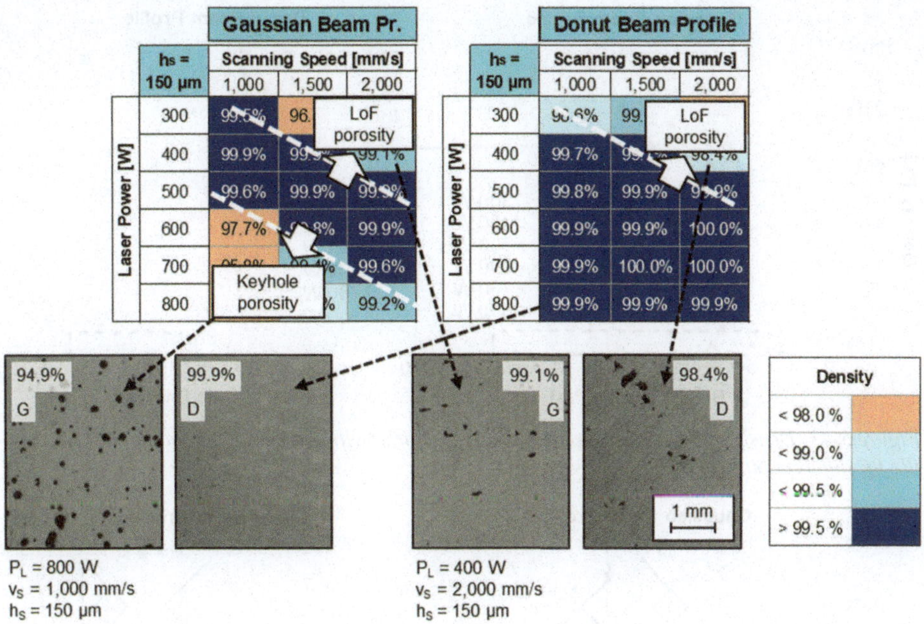

Figure 5-42: Occurrence of lack-of-fusion and keyhole porosity on process maps for the Gaussian (G) and donut (D) beam profile and a hatch distance hs = 150 µm for AlSi16Sc.

5.4.2 Effect on Build Rate

Below, the effect of the beam profile on the feasible build rate for AlSi16Sc is discussed.

Figure 5-41 shows the region of the highest build rate on the process maps. Both beam profiles successfully produce specimens with density > 99.5% with parameters of the maximum build rate tested. This is also demonstrated by Figure 5-43, that displays the specimen density for different build rates as defined by Equation 5.1 (page 95) [22]. However, it can be seen that the physical limits of the process are not reached in the experiments. Since it is not possible to do further experiments due to the non-availability of AlSi16Sc powder, and since it is presumed that the limits of the process will exceed the maximum laser power P_L (900 W) and scanning speed v_S (3,000 mm/s) of the available equipment, an estimation of the maximum possible build rates is done. The calculations follow the ones laid-out in Chapter 5.3.2 for AlSi10Mg.

Figure 5-44 illustrates the specimen density for different VED_{MOD} as defined by Equation 2.4 (page 16) for a hatch distance of h_S = 200 µm. It can be seen that keyhole welding occurs above values of $VED_{MOD\_MAX\_G}$ = 7.5 J/mm³ for the Gaussian beam profile and above $VED_{MOD\_MAX\_D}$ = 10.5 J/mm³ for the donut beam profile. The threshold VED_{MIN} for lack-of-fusion porosity is VED_{MIN} = 35 J/mm³ for both beam profiles and shown in Figure 5-36 (page 102).

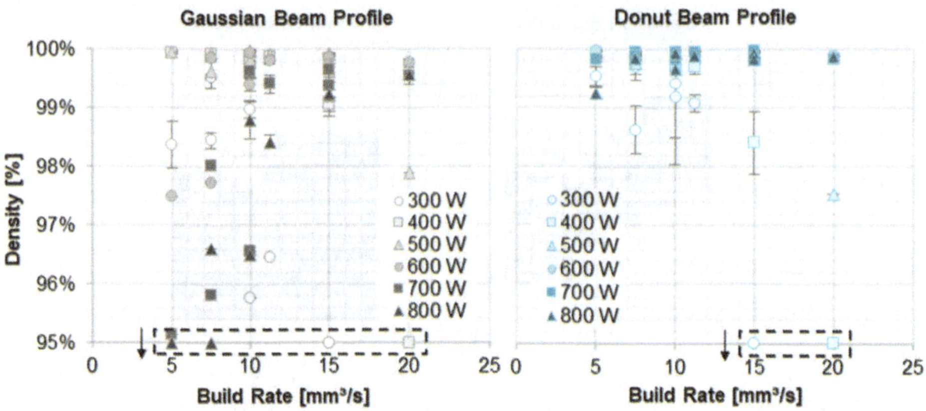

Figure 5-43: Density of AlSi16Sc specimens manufactured with the Gaussian and donut beam profile for different build rates.

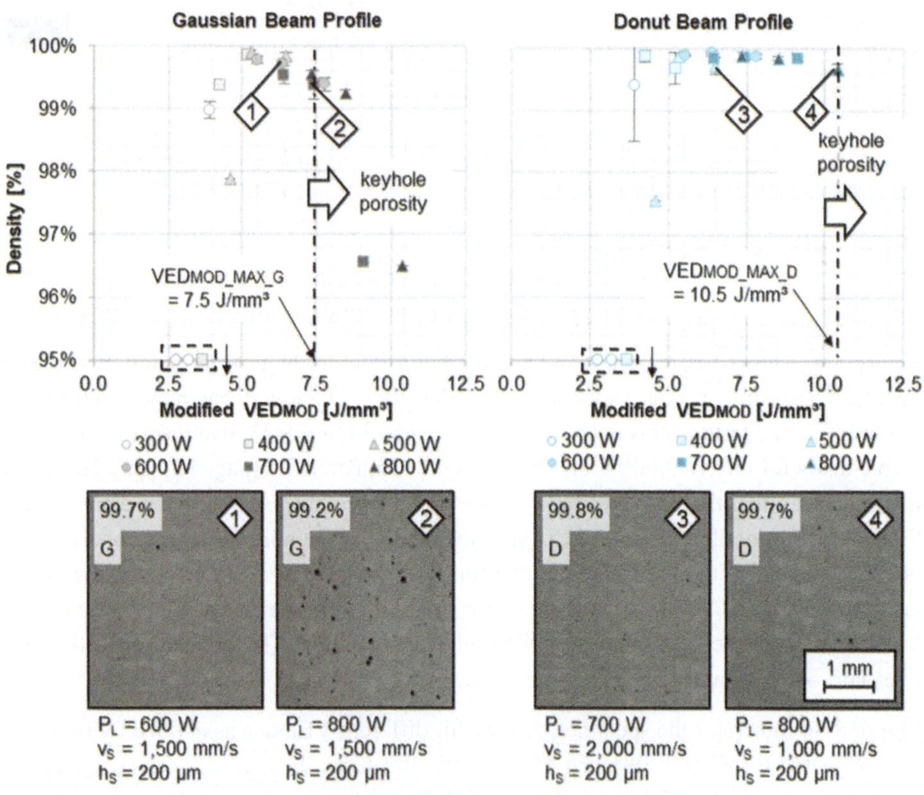

Figure 5-44: Density of specimens manufactured in AiSi16Sc for the Gaussian (G) and donut (D) beam profile and hatch distance of $h_S = 200\ \mu m$ for different VED_{MOD} with an indication of the threshold value for VED_{MOD_MAX} at which keyhole welding occurs.

Based on Equation 5.1 (page 95), Equation 5.3, and Equation 5.5 (both on page 98) and assuming the same constant material properties as for AlSi10Mg (see Chapter 5.3.2), the maximum feasible scanning speed v_{S_MAX}, build rate b_{MAX} and laser power P_{L_MAX} for the Gaussian and donut beam profiles are calculated in Equations 5.12 - 5.17.

$$v_{S\_MAX\_G2} = \frac{4 \cdot k \cdot d_f}{C_p \cdot \rho \cdot a_{fres}^2 \cdot h_l^2} \left(\frac{VED_{MOD\_MAX\_G}}{VED_{MIN\_G}} \right)^2 = 2{,}621 \; \frac{mm}{s} \tag{5.12}$$

$$v_{S\_MAX\_G2} = \frac{4 \cdot k \cdot d_f}{C_p \cdot \rho \cdot a_{fres}^2 \cdot h_l^2} \left(\frac{VED_{MOD\_MAX\_G}}{VED_{MIN\_G}} \right)^2 = 5{,}461 \; \frac{mm}{s} \tag{5.13}$$

$$b_{MAX\_G2} = v_{S\_MAX\_G} \cdot h_S \cdot h_l = 26 \; \frac{mm^3}{s} = 94 \; \frac{cm^3}{h} \tag{5.14}$$

$$b_{MAX\_D2} = v_{S\_MAX\_D} \cdot h_S \cdot h_l = 54 \; \frac{mm^3}{s} = 194 \; \frac{cm^3}{h} \tag{5.15}$$

$$P_{L\_MAX\_G2} = v_{S\_MAX\_G} \cdot h_S \cdot h_l \cdot VED_{MIN} = 917 \; W \tag{5.16}$$

$$P_{L\_MAX\_D2} = v_{S\_MAX\_D} \cdot h_S \cdot h_l \cdot VED_{MIN} = 2{,}041 \; W \tag{5.17}$$

$v_{S\_MAX\_G2}$: Maximum feasible scanning speed for Gaussian beam profile for AlSi16Sc [mm/s]
$v_{S\_MAX\_D2}$: Maximum feasible scanning speed for donut beam profile for AlSi16Sc [mm/s]
b_{MAX_G2}: Maximum feasible build rate for Gaussian beam profile for AlSi16Sc [mm^3/s]
b_{MAX_D2}: Maximum feasible build rate for donut beam profile for AlSi16Sc [mm^3/s]
$P_{L\_MAX\_G2}$: Maximum feasible laser power for Gaussian beam profile for AlSi16Sc [W]
$P_{L\_MAX\_D2}$: Maximum feasible laser power for donut beam profile for AlSi16Sc [W]
$VED_{MOD\_MAX\_G2}$: Maximum *modified* VED_{MOD} for Gaussian beam profile for AlSi16Sc (7.5 J/mm^3)
$VED_{MOD\_MAX\_D2}$: Maximum *modified* VED_{MOD} for donut beam profile for AlSi16Sc (10.5 J/mm^3)
VED_{MIN_2}: Minimum VED for Gaussian and donut beam profile for AlSi16Sc (35 J/mm^3)
k: Thermal conductivity (100 W/(m·K))
C_p: Specific heat capacity (1,030 J/kgK)
a_{fres}: Absorption coefficient (0.4)
h_S: Hatch distance (200 µm)
d_f: Focus diameter of laser (140 µm)
ρ: Density of material (2,400 kg/m^3)
h_l: Layer thickness (50 µm)

The results show that the maximum feasible build rate is b_{MAX_G} = 26 mm^3/s for the Gaussian beam profile and b_{MAX_D} = 54 mm^3/s for the donut beam profile. Therefore, it is estimated that the donut beam profile offers the potential to increase the build rate by more than 100% for AlSi16Sc. Compared to the results for AlSi10Mg, the maximum feasible build rate with the Gaussian beam profile is 30% lower (b_{MAX_D} = 37 mm^3/s for AlSi10Mg and 26 mm^3/s AlSi16Sc), while it is < 10% lower in the case of donut beam profile (b_{MAX_D} = 59 mm^3/s for AlSi10Mg and 54 mm^3/s AlSi16Sc). It might be surprising at first sight that the difference between the materials is not as significant for the donut beam profile, but this can be explained by the circumstance that the maximum feasible energy input of $VED_{MOD\_MAX\_D}$ = 11 J/mm^3 for AlSi10Mg is a lower boundary. Therefore, the correct value for $VED_{MOD\_MAX\_D}$ is might be higher, which will lead to higher maximum build rates for AlSi10Mg. Furthermore, it has to be mentioned that the calculations performed in this chapter are done with the material properties of AlSi10Mg as they are not available for AlSi16Sc. The results will be slightly different with the correct material properties of AlSi16Sc. However, it is assumed that the calculations provide a reasonable estimation of the feasible build rates.

The circumstance that the maximum scanning speed and build rate can be increased by more than 100% with the donut beam profile can be seen as a significant advantage that leads to relevant cost reductions of the SLM process for AlSi16Sc.

5.4.3 Summary of Results

To investigate the effect of the beam profile on the multi-laser SLM process for AlSi16Sc, density cubes were manufactured and the porosity was analyzed by micro-sections. Based on the results, the effect of the beam profile on the process stability and build rate was discussed.

It can be concluded that there is a significant effect of the beam profile on the processing of AlSi16Sc by SLM. Although both beam profiles can manufacture highly dense parts (> 99.5%), the porosity is higher for the specimens manufactured with the Gaussian beam profile for almost all parameters. At low VEDs (< 35 J/mm^3), lack-of-fusion porosity is present for both beam profiles, but only for the Gaussian beam profile, a relevant increase in spherical porosity occurs at VEDs > 65 J/mm^3. Consequently, the process window that produces densities > 99.5% is significantly larger for the donut beam profile. Compared to the results obtained for AlSi10Mg, the process window is similar for the donut beam profile, but considerably smaller for the Gaussian beam profile. This suggests that the donut beam profile offers a particular advantage for processing materials that are difficult to process in SLM. The primary cause for the increase in spherical porosity observed in the specimens manufactured with the Gaussian beam profile, compared to AlSi10Mg, is seen in the lower hydrogen solubility, the higher melt velocities, which increase the gas entrapment, and the amplified melt pool instabilities resulting from the reduced viscosity of AlSi16Sc.

In addition, the beam profile has a considerable effect on the productivity of the process. Since it was not possible to test the physical limits of the process with the available equipment and due to the non-availability of a sufficient amount of AlSi16Sc powder, calculations on the maximum feasible build rates were done, based on the experimental results. The outcomes reveal that an increase in build rate of more than 100% seems possible with the donut beam profile compared to the Gaussian beam profile. However, experimental investigations are necessary in the future to verify these results.

6 Industrial Relevance of Results

In this chapter, the industrial relevance of the results laid out in Chapter 5 is discussed (Chapter 6.1). Furthermore, an outlook on the effect of potential future process enhancements is provided (Chapter 6.2).

6.1 Effect on Manufacturing Quality, Time and Costs

The significant challenges the SLM technology faces in the industry today are high costs and issues with part quality as well as reproducibility [202, 240]. The high costs are directly related to the low machine productivity in combination with high investment costs (challenges: cost & time). The issues in quality are related to instabilities of the process and non-controlled process parameters that lead to long-term process shifts (challenge: quality) [139, 202]. The numerical and experimental results demonstrate several advantages the donut beam profile offers with which to attenuate the mentioned challenges.

The **quality of the process** is improved due to several advancements. First, the overall amount of porosity is reduced, especially at higher energy inputs. This is achieved due to a reduction in vaporization of (on average) more than 40% and due to a more homogeneous melt flow. Second, the amount of spatters, which can produce defects when they fall onto the powder bed or interfere with the laser beam, is reduced by > 35%. In combination with the larger process window, this makes the process more robust against variations and (potentially) increases the long-term stability. However, further investigations are necessary to verify the latter.

The **manufacturing time** is reduced thanks to an increase in the maximum feasible build rate of 60% by the donut beam profile for AlSi10Mg. To estimate the effect of the higher build rate on the overall productivity of the process (considering jump vectors and recoating time), exemplary calculations are done in the next paragraph for selected use cases.

The **manufacturing costs** are reduced since they depend, to a high degree, on the manufacturing time [10, 240]. To quantify this, the effect of the advanced process on the manufacturing costs and time for three use cases are calculated. The following use cases have been selected (see Figure 6-1):

- An optimized AM *heat sink* for an LED light developed by Fraunhofer IAPT [115]. The optimized heat sink offers a weight saving and reduction in thermal resistance of more than 70% in comparison to the conventional design.
- An *FCRC bracket,* developed by Airbus Operations GmbH and LZN Laser Zentrum Nord GmbH (now Fraunhofer IAPT) in 2013, which was the first flying 3D printed metal part on a commercial aircraft in 2014 [6, 201]. The bracket is one out of several brackets that fixates the Flight Crew Rest Compartment (FCRC) to the primary structure of an Airbus A350.

T. M. Wischeropp, *Advancement of Selective Laser Melting by Laser Beam Shaping*,
Light Engineering für die Praxis, https://doi.org/10.1007/978-3-662-64585-7_6

- A *Valve block*, developed by Fraunhofer IAPT, which offers a weight saving of 80% and reduction of pressure loss of 40%, compared to the conventional valve design [60].

For this analysis, it is assumed that all parts are manufactured in AlSi10Mg, although the FCRC bracket (TiAl6V4) and valve block (SS316L) are usually manufactured in other materials.

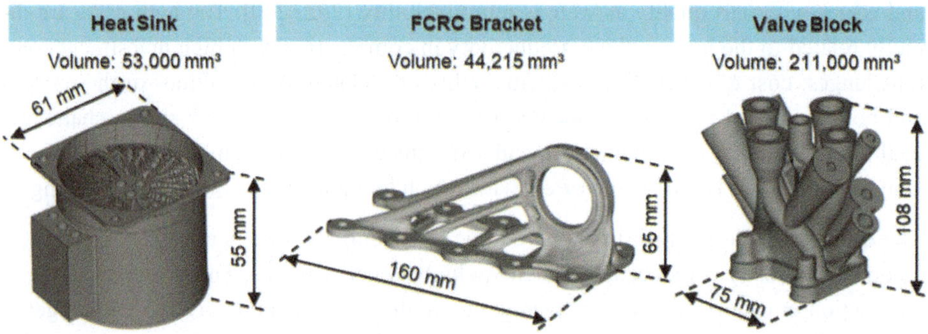

Figure 6-1: Use cases for calculation of business cases.

The manufacturing time and costs are calculated for the serial production of the use cases on an SLM 500 HL Quad Laser machine, as shown in Figure 6-2. It is assumed that the advanced process can be directly transferred from the test-rig to the industrial machine and that no limits in available laser power and scanning speeds exist. Furthermore, the costs of labor and post-processing (e.g., heat treatment or surface finishing) are not considered since they are not influenced by the results obtained in this thesis and since they are highly part-dependent.

For the calculations, an hourly machine rate of 71 €/h is used, based on the calculations shown in Appendix A.9. Since the costs for the beam shaping equipment (< 2,000 €) are negligible, compared to the overall machine costs, the same hourly machine rate is used for the set-up with and without beam shaping. Table 6-1 lists the essential data used for the calculations.

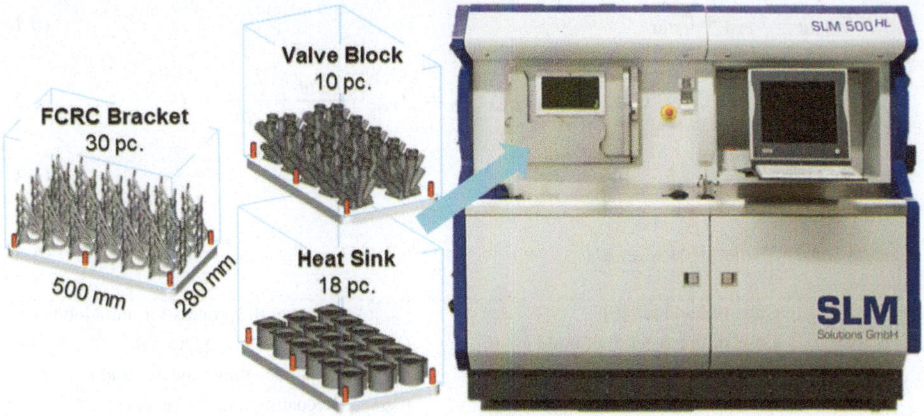

Figure 6-2: Positioning of use cases on the build platform of an SLM 500 HL Quad Laser machine from SLM Solutions GmbH (now SLM Solutions Group AG)

Table 6-1: Relevant data used for calculation of the manufacturing time and costs.

Data		Source
Recoating time per layer t_{rec}:	6 s	measured at machine
Build rate b_{GST} of standard parameters for AlSi10Mg:	4·12.5 mm³/s	Table 4-3
Max. build rate b_{MAX_G} with Gaussian beam profile:	4·37 mm³/s	Chapter 5.3.2
Max. build rate b_{MAX_D} with donut beam profile:	4·59 mm³/s	Chapter 5.3.2
Leveling height of build platform h_l:	0.05 mm	Table 4-1
Costs of AlSi10Mg powder C_{Mat_kg}:	50 €/kg	av. market price
Hourly machine rate C_h:	71 €/h	see Appendix A.9
Volume and number of parts per build job:	Figure 6-1 and Figure 6-2	
Building height of build job h_{build}:	Figure 6-1	

As previously discussed, the costs of SLM parts are mainly driven by the process and material costs. They can be calculated by Equation 6.1 (neglecting labor costs) [188, 240]. The costs of the process C_{Pro} are subject to the processing time t_{build} multiplied by the hourly machine rate C_h (see Equation 6.2). It is assumed that the *actual* build rate br is 20% lower than the build rate b due to jump-vectors and laser-scanner-delays as shown in Equation 6.3. The circumstance that this factor is not constant and depends on the geometry of the part is neglected for the matter of simplification (e.g., the amount of jump vectors in relation to the volume of the part increases for fine structures). The costs of the material C_{Mat} are defined by the total volume of the build job V_{build} multiplied by the density ρ and costs C_{Mat_kg} of the powder material (see Equation 6.4).

$$C_{build} = C_{Pro} + C_{Mat} \tag{6.1}$$

$$C_{Pro} = t_{build} \cdot C_h = (t_{rec\_t} + t_{melt\_t}) \cdot C_h = (t_{rec} \cdot \frac{h_{build}}{h_l} + \frac{V_{build}}{br}) \cdot C_h \tag{6.2}$$

$$br = K_j \cdot b_{MAX\_G,D} \tag{6.3}$$

$$C_{Mat} = V_{build} \cdot \rho \cdot C_{Mat\_kg} \tag{6.4}$$

C_{build}:	Costs of build job [€]	C_{Pro}:	Costs of the process for build job [€]
C_{Mat}:	Costs of the material of build job [€]	C_h:	Hourly machine rate [€/h]
t_{build}:	Total build time of build job [s]	t_{rec_t}:	Total recoating time of build job [s]
t_{melt_t}:	Total melting time of build job [s]	t_{rec}:	Recoating time per layer [s]
h_{build}:	Height of build job [mm]	h_l:	Leveling height of build platform [mm]
V_{build}:	Total volume of parts in build job [mm³]	br:	Actual build rate [mm³/s]
K_j:	Factor for jump vectors and delays (0.8) [1]	C_{Mat_kg}:	Costs for material per kg [€/kg]
ρ:	Density of material [kg/m³]	b_{MAX_G}:	Max. build rate for G. profile [mm³/s]
b_{MAX_D}:	Max. build rate for donut profile [mm³/s]		

The manufacturing time and costs for the use cases are calculated, based on Equations 6.1-6.4 and data provided in Table 6-1. The calculations are done for three different scenarios:

- manufacturing with today's **standard process parameters** (reference parameters) with Gaussian beam profile and focus diameter of 85 µm,
- manufacturing with **Gaussian beam profile** and focus diameter of 140 µm with the maximum feasible build rate b_{MAX_G},
- manufacturing with **donut beam profile** and focus diameter of 140 µm with the maximum feasible build rate b_{MAX_D}.

Figure 6-3 shows the results of the calculations. It can be seen that there is a significant effect of the beam profile as well as beam size on the manufacturing time and costs and that the effect depends on the part volume and geometry. Compared to the reference parameters, the manufacturing costs can be reduced by up to 49% with an increase in beam diameter (85 µm to 140 µm) and an additional 15% by the donut beam profile. Furthermore, the manufacturing time can be reduced by up to 58% by an increase in beam diameter and an additional ≥ 20% by the donut beam profile. It must be noted that even further manufacturing time and cost reductions might be feasible with the donut beam profile as the build rate that is used for the calculations is (potentially) not the maximum feasible build rate (see Chapter 5.3.2).

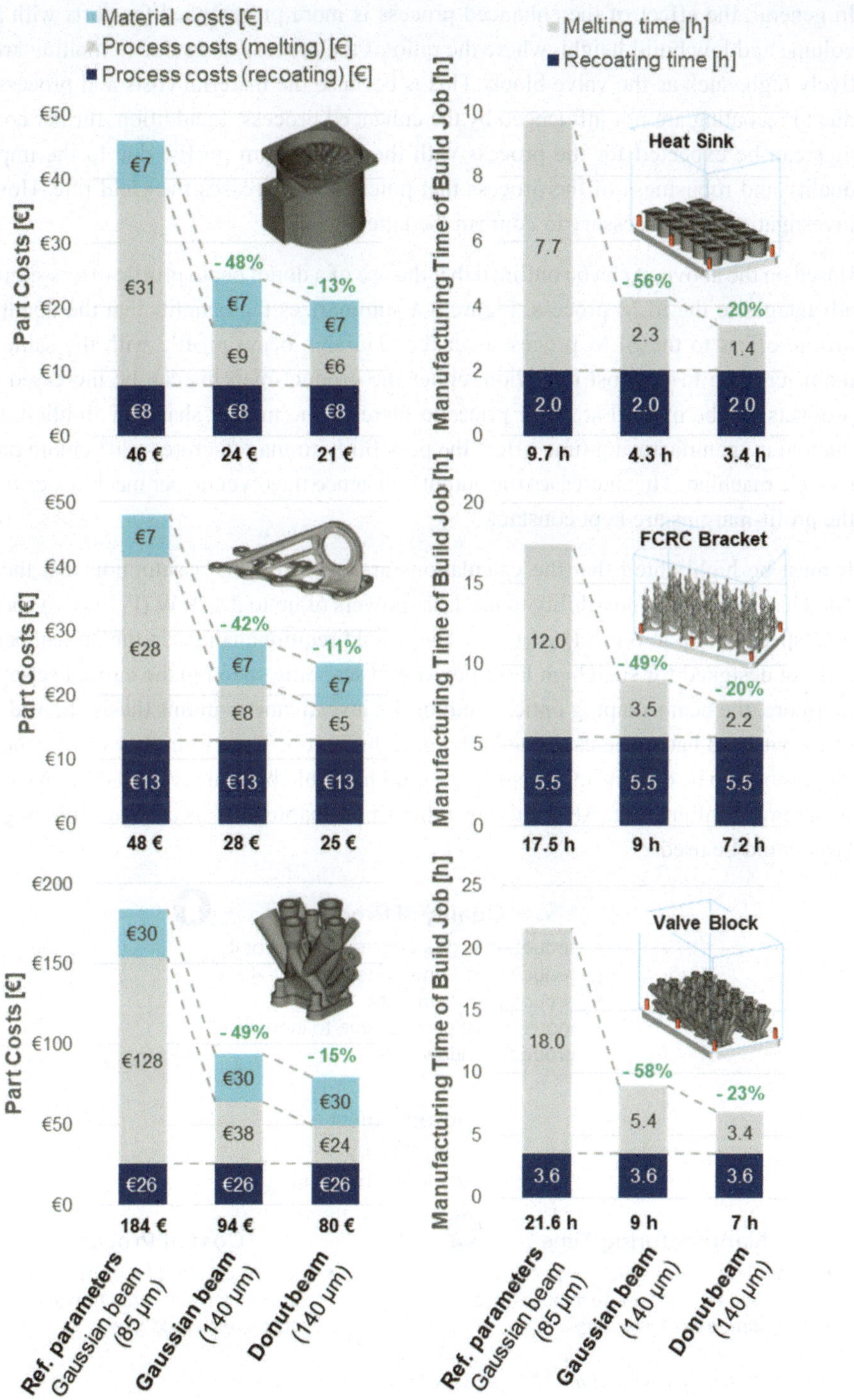

Figure 6-3: Effect of the enhanced SLM process on the manufacturing time and costs for selected use cases manufactured in AlSi10Mg on an SLM 500 HL.

In general, the effect of the enhanced process is more pronounced for parts with a high volume and low build height, where the ratio of the process costs due to melting are relatively high, such as the valve block. This is because the material costs and process costs due to recoating are not influenced by the enhanced process. In addition, further cost savings can be expected for the process with the donut beam profile due to the improved quality and robustness of the process that potentially increases the yield rate. However, investigations are necessary to confirm the latter.

Based on the above, it can be outlined that the use of a donut beam profile offers significant advantages to the SLM process. Figure 6-4 summarizes the benefits that the donut beam profile offers to the SLM process over the Gaussian beam profile with the same focus diameter. Due to the cost reduction, either the product-margins can be increased or the products can be offered at lower prices to increase the market share. In addition, the reduction in manufacturing time offers the possibility to manufacture > 20% more parts on a single machine. This increases the output and hence the revenue per machine, even when the profit-margins are kept constant.

It must be highlighted that the calculations are based on the assumption that the SLM 500 HL provides the possibility to use laser powers of up to 2,079 W ($P_{L\_MAX\_D}$) and scanning speeds of up to 6 m/s ($v_{S\_MAX\_D}$). This would require changes in the optical bench as it is not designed for such high laser power and scanning speed in the current set-up. Furthermore, the beam shaping optics used for the investigations in this thesis showed a high sensitivity and had to be calibrated after each build job. Therefore, they need to be made more robust to be used in an industrial set-up. This could be achieved by additional cooling or automated alignment. Alternatively, other beam shaping optics or adjustable ring mode laser could be used.

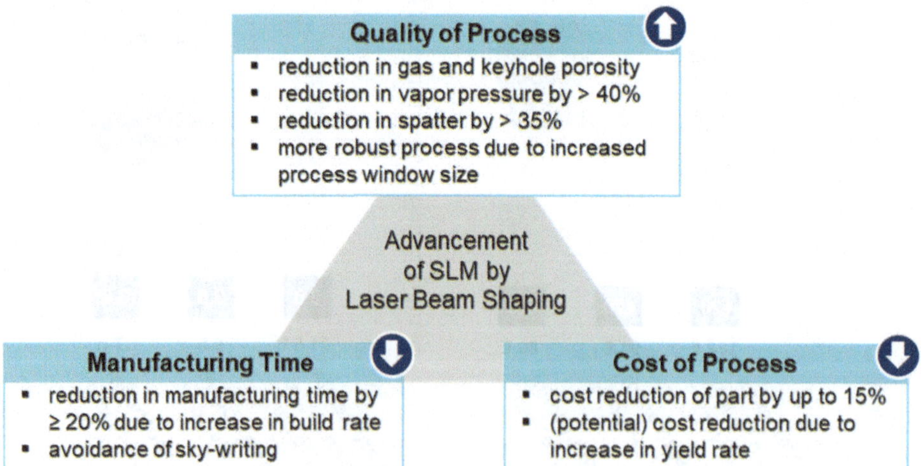

Figure 6-4: Advancement of the SLM process by the donut beam profile.

6.2 Effect of (Potential) Future Enhancements

From the results presented in Figure 6-3 (previous chapter), it can be seen that the highest share of the manufacturing time and costs are related to the melting of the material when utilizing the reference parameters (Gaussian beam with 85 µm diameter). With the enhanced process achieved by the donut beam profile, these costs become the lowest of the three main cost drivers (material, process costs for recoating, process costs for melting) for all use cases. To estimate the effect of future process enhancements on the business case, calculations of the manufacturing time and costs are done for two additional scenarios:

- manufacturing with **donut beam profile** with a maximum feasible build rate that is 50% higher than the maximum build rate b_{MAX_D} (see Table 6-1),
- manufacturing with **donut beam profile** with a maximum feasible build rate that is 50% higher than the maximum build rate b_{MAX_D} and a layer thickness that is increased by 100%.

The calculations are done for one of the use cases (heat sink). As the material costs are not influenced by the process itself, they are kept constant in the calculation.

Figure 6-5 depicts the effect of the (potential) process enhancements on the manufacturing costs and time for the selected use case. An additional increase in build rate by 50% (e.g., by additional laser sources or higher scanning speeds) would lead to additional manufacturing time and costs savings of ~10%. However, if the higher build rates are combined with a higher layer thickness, the manufacturing time and costs for recoating are lowered as well. This would lead to cost savings of almost 30% and a reduction in manufacturing time of ~40% for the selected use case. This demonstrates that future process enhancements should not solely focus on higher scanning speeds and hatch distances, but also on an increase in layer thicknesses to enhance the industrial relevance of the outcomes.

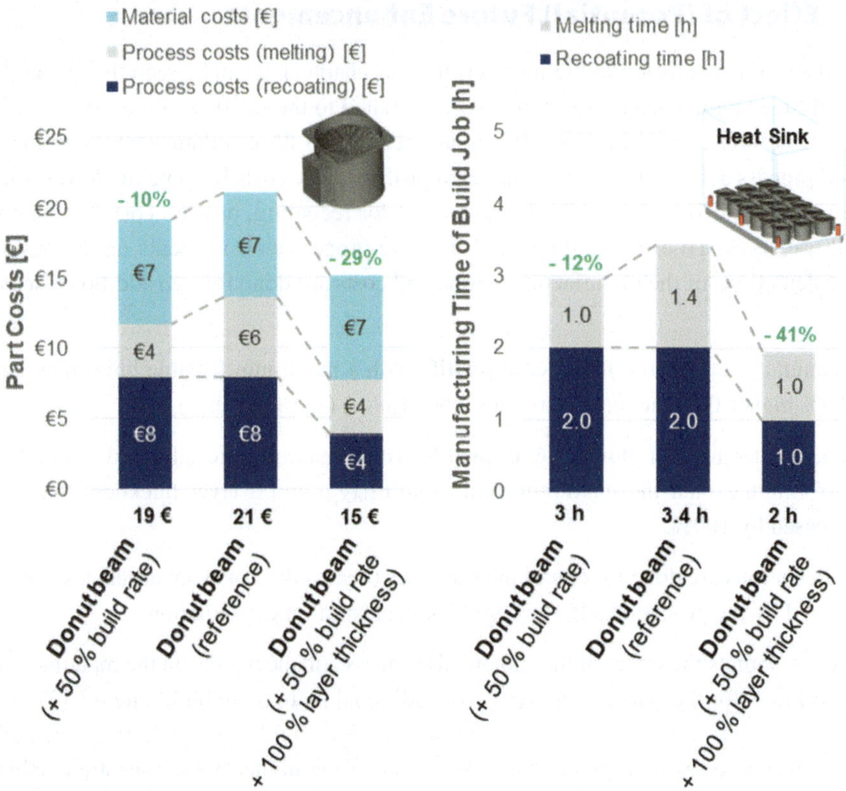

Figure 6-5: Effect of potential future process enhancements on the manufacturing time and costs of a selected use case.

7 Summary and Outlook

Despite the manifold advantages Selective Laser Melting (SLM) offers to manufacturing companies, the industrial application is still limited, especially in serial production. Major challenges that impede the industrial upscaling are the high costs as well as issues in the quality and reproducibility of the process. Therefore, it was the goal of this thesis to investigate the potential an optimized laser beam profile offers to overcome these challenges. For this, a numerical micro-scale model was developed to gain insights into the effect of the beam profile on relevant melt pool characteristics that are difficult to investigate experimentally. Furthermore, single track and density cube experiments were performed with AlSi10Mg and AlSi16Sc on a modified SLM machine using customized beam shaping optics. The experiments were done with a Gaussian beam profile that is used in current SLM machines and a donut beam profile with a more homogeneous energy distribution. The results obtained for both beam profiles were compared and explained with the help of the numerical outcomes. Finally, the industrial relevance of the results was discussed.

The results of the numerical investigations reveal that the melt pool characteristics are considerably affected by the beam profile. Due to the reduced intensity and more homogeneous energy distribution of the donut beam profile, the vaporization is reduced by more than 40%, on average. Furthermore, the thermo-capillary flow is attenuated due to the lower temperature gradients on the melt pool surface. This leads to melt velocities that are, on average, 20% lower compared to the melt velocities caused by the Gaussian beam profile. Consequently, the melt pools are similar in width, but shallower (~20%) and shorter (~10%) for the donut beam profile. In addition, the melt pool caused by the donut beam profile shows fewer fluctuations than the one caused by the Gaussian beam profile, especially at high energy inputs. This demonstrates a stabler melting process.

The outcomes of the single-track experiments affirm the results obtained from the simulation. The single tracks manufactured with the donut beam profile show fewer irregularities and are slightly more homogeneous due to a stabler melting process. Moreover, the desired symmetric cross-sections of the single tracks are obtained in a wider process window for the donut beam profile.

The effect of the beam profile on the multi-layer SLM process for AlSi10Mg is considerably high. Although both beam profiles can manufacture highly dense parts in a wide process window, the overall porosity is lower when the donut beam profile is used. This difference is particularly significant at high energy inputs, where an increase in spherical porosity is observed for the Gaussian beam profile that comes at least partly from collapsing keyholes. Contrary to this, fully dense parts are produced even at the highest energy input tested with the donut beam profile. Due to the higher feasible energy input, an increase in build rate of 60% seems possible with the donut beam profile, based on theoretical analyses. In addition, the amount of spatters is reduced by >35%, which reduces the risk of defects. The major reason for the observed differences is seen in the reduced vapor

T. M. Wischeropp, *Advancement of Selective Laser Melting by Laser Beam Shaping*,
Light Engineering für die Praxis, https://doi.org/10.1007/978-3-662-64585-7_7

pressure and lower melt velocities caused by the donut beam profile that leads to a stabler melting process.

The effect of the beam profile on the processing of the high strength AlSi16Sc alloy is even more pronounced than for AlSi10Mg. For the Gaussian beam profile, a significant increase in spherical porosity (gas and keyhole porosity), compared to AlSi10Mg, occurs at medium to high energy inputs. The increase in porosity comes from two effects: First, the lower hydrogen solubility of AlSi16Sc increases the amount of gas bubbles that nucleate during solidification, and the more pronounced thermo-capillary flow increases the risk of their entrapment in the melt pool. Second, an increase in keyhole instability is expected due to the lower viscosity of the alloy. Consequently, the process window size is reduced considerably. In contrast to this, the process window size is comparable to the one of AlSi10Mg for the donut beam profile and process parameters tested. Due to the higher feasible energy input, an increase in build rate of more than 100% seems possible with a donut beam profile for AlSi16Sc, compared to the Gaussian beam profile. Furthermore, the larger process window of the donut beam profile demonstrates a more robust, stabler process.

In summary, the results show that there is a significant effect of the laser beam profile on the SLM process and that a donut beam profile offers several advantages over a standard Gaussian beam profile. First, it increases the quality of the parts by reducing the overall porosity, decreases the amount of spatters and increases the robustness of the process due to an increase in process window size. Second, it reduces the manufacturing time by $\geq 20\%$ (for AlSi10Mg) due to an increase in the build rate and third, it reduces costs due to the increase in the build rate by up to 15%.

The presented work is the first study investigating the effect of the beam profile on the process stability (feasible energy input, process window size, defect generation) and productivity (build rate) in such detail. Although the results are promising, a significant amount of future work must be performed to fully exploit the potential of beam shaping for SLM and to establish it in the industry. Foremost, it is recommended that more experimental work is carried out to confirm the physical limits of the process in terms of energy input and maximum build rate. For this, experiments with higher laser powers, scanning speeds and layer thicknesses are recommended. Furthermore, the effect of the beam profile on other important characteristics such as microstructure, surface roughness and, ultimately, the mechanical properties have to be studied. Moreover, the investigation of the potential beam shaping offers for processing other materials such as titanium, steel or nickel-base alloys is recommended. Based on the results obtained for AlSi16Sc it could be expedient, in particular, to utilize beam shaping for materials that are difficult to process by SLM. Lastly, the study of additional beam profiles such as top-hat or alternative shapes of the donut beam should be considered and the process should be validated for complex three-dimensional parts.

In addition to the above, future work is necessary to transfer the potential of beam shaping to industrial SLM machines. The equipment used for the investigations in this thesis was very sensitive and had to be calibrated after each build job. Therefore, other beam shaping techniques are recommended and should be investigated. In this context, adjustable ring-mode lasers that can generate different beam profiles and change the profile "on the fly" without the need for additional beam shaping optics are of particular interest. Moreover, investigations on the effect of the beam profile on the long-term stability of the SLM process should be conducted, as this is of high importance for industrial applications.

References

[1] Aboulkhair, N. T., Everitt, N. M., Ashcroft, I., and Tuck, C. 2014. Reducing porosity in AlSi10Mg parts processed by selective laser melting. *Additive Manufacturing* 1-4, 77–86.

[2] Aboulkhair, N. T., Maskery, I., Tuck, C., Ashcroft, I., and Everitt, N. 2015. Nano-hardness and microstructure of selective laser melted AlSi10Mg scan tracks. In *Industrial Laser Applications Symposium (ILAS 2015)*. SPIE Proceedings. SPIE, 965702. DOI=10.1117/12.2190015.

[3] Aboulkhair, N. T., Maskery, I., Tuck, C., Ashcroft, I., and Everitt, N. M. 2016. On the formation of AlSi10Mg single tracks and layers in selective laser melting. Microstructure and nano-mechanical properties. *Journal of Materials Processing Technology* 230, 88–98.

[4] Aboulkhair, N. T., Simonelli, M., Parry, L., Ashcroft, I., Tuck, C., and Hague, R. 2019. 3D printing of Aluminium alloys: Additive Manufacturing of Aluminium alloys using selective laser melting. *Progress in Materials Science* 106, 100578.

[5] Afazov, S., Denmark, W. A. D., Toralles, B. L., Holloway, A., and Yaghi, A. 2017. Distortion prediction and compensation in selective laser melting. *Additive Manufacturing* 17, 15–22.

[6] Airbus SE. 2018. *3D Printing. Decarbonisation meets digitalisation*. https://www.airbus.com/public-affairs/brussels/our-topics/innovation/3d-printing.html. Accessed 3 October 2019.

[7] Akashev, L. A. and Kononenko, V. I. 2001. Optical Properties of Liquid Aluminum and Al-Ce Alloy. *High Temperature* 39, 3, 412–415.

[8] American Welding Society. Influence of Fluid Convection on Weld Pool Formation in Laser Cladding. *Welding Journal,* August 2014, Vol. 93.

[9] Ammer, R., Markl, M., Jüchter, V., Körner, C., and Rüde, U. 2014. Validation Experiments for LBM Simulations of Electron Beam Melting. *International Journal of Modern Physics C* 25.

[10] Ampower GmbH & Co. KG. 2019. *AMPOWER Report 2019*. AM Power GmbH & Co KG, Hamburg.

[11] Antony K. and Arviazhagan N. 2015. Studies on Energy Penetration and Marangoni Effect during Laser Melting Process. *Journal of Engineering Science and Technology* 10, 4, 509–525.

[12] Anyalebechi, P. Analysis of the Effects of Alloying Elements on Hydrogen Solubility in Liquid Aluminium Alloys. *Scripta Metallurgica et Materialia* 1995, Vol. 33, No. 8, pp. 1209-1216.

T. M. Wischeropp, *Advancement of Selective Laser Melting by Laser Beam Shaping*, Light Engineering für die Praxis, https://doi.org/10.1007/978-3-662-64585-7

[13] Arnold, D. N. 2015. Stability, consistency, and convergence of numerical discretizations. *Encyclopedia of Applied and Computational Mathematics*.

[14] Attar, E. 2011. *Simulation of Selective Electron Beam Melting Process. Lattice Boltzmann modeling of beam based additive manufacturing processes*. VDM Verlag Dr. Müller.

[15] Attar, E. and Körner, C. 2011. Lattice Boltzmann model for thermal free surface flows with liquid–solid phase transition. *International Journal of Heat and Fluid Flow* 32, 1, 156–163.

[16] Attar, H., Calin, M., Zhang, L. C., Scudino, S., and Eckert, J. 2014. Manufacture by selective laser melting and mechanical behavior of commercially pure titanium. *Materials Science and Engineering: A* 593, 170–177.

[17] Awd, M., Siddique, S., Johannsen, J., Emmelmann, C., and Walther, F. 2019. Very high-cycle fatigue properties and microstructural damagemechanisms of selective laser melted AlSi10Mg alloy. *International Journal of Fatigue*.

[18] Bäckerud, L., Chai, G., and Tamminen, J. 1990. *Foundry alloys*. Solidification Characteristics of Aluminum Alloys 2. Skanaluminium, Oslo.

[19] Basoalto, H. C., Panwisawas, C., Sovani, Y., Anderson, M. J., Turner, R. P., Saunders, B., and Brooks, J. W. 2018. A computational study on the three-dimensional printability of precipitate-strengthened nickel-based superalloys. *Proceedings. Mathematical, physical, and engineering sciences* 474, 2220, 20180295.

[20] Bayat, M., Thanki, A., Mohanty, S., Witvrouw, A., Yang, S., Thorborg, J., Tiedje, N. S., and Hattel, J. H. 2019. Keyhole-induced porosities in Laser-based Powder Bed Fusion (L-PBF) of Ti6Al4V: High-fidelity modelling and experimental validation. *Additive Manufacturing* 30, 100835.

[21] Bergström, D. 2008. *The Absorption of Laser Light by Rough Metal Surfaces*. Doctoral Thesis, Luleå University of Technology.

[22] Brandt, M., Ed. 2016. *Laser Additive Manufacturing. Materials, design, technologies, and applications*. Woodhead Publishing series in electronic and optical materials 88. Woodhead Publishing, Oxford.

[23] Brodova, I. G., Popel, P. S., and Eskin, G. I. 2001. *Liquid Metal Processing. Applications to Aluminium Alloy Production*. Advances in Metallic Alloys. Taylor & Francis.

[24] Chen, C., Gu, D., Dai, D., Du, L., Wang, R., Ma, C., and Xia, M. 2019. Laser additive manufacturing of layered TiB2/Ti6Al4V multi-material parts: Understanding thermal behavior evolution. *Optics & Laser Technology* 119, 105666.

[25] Chen, Z., Xiang, Y., Wei, Z., Wei, P., Lu, B., Zhang, L., and Du, J. 2018. Thermal dynamic behavior during selective laser melting of K418 superalloy: numerical simulation and experimental verification. *Appl. Phys. A* 124, 4, 168.

[26] Cheng, B., Li, X., Tuffile, C., Ilin, A., Willeck, H., and Hartel, U. 2018. Multi-Physics Modeling of Single Track Scanning in Selective Laser Melting: Powder Compaction Effect. *Proceedings of the 29th Annual International Solid Freeform Fabrication Symposium*, 1887–1902.

[27] Cheng, B., Loeber, L., Willeck, H., Hartel, U., and Tuffile, C. 2019. Computational Investigation of Melt Pool Process Dynamics and Pore Formation in Laser Powder Bed Fusion. *Journal of Material Engineering and Performance* 28, 11, 6565–6578.

[28] Cheng, B. and Tuffile, C. Numerical Study of Porosity Formation With Implementation of Laser Multiple Reflection in Selective Laser Melting. In *ASME 2019 14th International Manufacturing Science and Engineering Conference*.

[29] Childs, T. H. C., Hauser, C., and Badrossamay, M. 2005. Selective laser sintering (melting) of stainless and tool steel powders. Experiments and modelling. *Proceedings of the Institution of Mechanical Engineers, Part B: Journal of Engineering Manufacture* 219, 4, 339–357.

[30] Childs, T.H.C., Hauser, C., and Badrossamay, M. 2004. Mapping and Modelling Single Scan Track Formation in Direct Metal Selective Laser Melting. *CIRP Annals - Manufacturing Technology* 53, 1, 191–194.

[31] Chivel, Y. Ablation phenomena and instabilities under laser melting of powder layers. In *8th International Conference on Photonic Technologies Lane 2014. DOI*=10.1016/S1875-3892(14)00349-6.

[32] Cho, J.-H. and Na, S.-J. 2006. Implementation of real-time multiple reflection and Fresnel absorption of laser beam in keyhole. *J. Phys. D: Appl. Phys.* 39, 24, 5372–5378.

[33] Cloots, M. 2017. *Empirische und simulative Studie über die Verarbeitbarkeit von IN738LC mittels SLM.* Doctoral Thesis. ETH Zürich.

[34] Cloots, M., Uggowitzer, P. J., and Wegener, K. 2016. Investigations on the microstructure and crack formation of IN738LC samples processed by selective laser melting using Gaussian and doughnut profiles. *Materials & Design* 89, 770–784.

[35] Cunningham, R., Zhao, C., Parab, N., Kantzos, C., Pauza, J., Fezzaa, K., Sun, T., and Rollett, A. D. 2019. Keyhole threshold and morphology in laser melting revealed by ultrahigh-speed x-ray imaging. *Science (New York, N.Y.)* 363, 6429, 849–852.

[36] Dai and Gu, D. 2015. Tailoring surface quality through mass and momentum transfer modeling using a volume of fluid method in selective laser melting of TiC/AlSi10Mg powder. *International Journal of Machine Tools and Manufacture* 88, 95–107.

[37] Dai, D. and Gu, D. 2014. Thermal behavior and densification mechanism during selective laser melting of copper matrix composites. Simulation and experiments. *Materials & Design* 55, 482–491.

[38] Deibler, L., Brown, B., Finrock, C., and Carroll, J., Coker, E. 2017. Effects of AlSi10Mg feedstock condition on part properties. *Annual International Solid Freeform Fabrication Symposium.*

[39] Deng, S., Wang, H.-P., Lu, F., Solomon, J., and Carlson, B. E. 2019. Investigation of spatter occurrence in remote laser spiral welding of zinc-coated steels. *International Journal of Heat and Mass Transfer* 140, 269–280.

[40] Di, W., Yongqiang, Y., Xubin, S., and Yonghua, C. 2012. Study on energy input and its influences on single-track,multi-track, and multi-layer in SLM. *The International Journal of Advanced Manufacturing Technology* 58, 9-12, 1189–1199.

[41] Dianat, M., Skarysz, M., and Garmory, A. 2017. A Coupled Level Set and Volume of Fluid method for automotive exterior water management applications. *International Journal of Multiphase Flow* 91, 19–38.

[42] Diaz del Castillo, J. M. 2017. *Development of an FEM-model for a quantitative evaluation of the melt pool size generated by different laser beam intensity profiles in Selective Laser Melting.* Master Thesis, TU Hamburg-Harburg.

[43] Dickey, F. M. and Holswade, S. C. 2000. *Laser Beam Shaping. Theory and Techniques.* Marcel Dekker, Inc., New York, Basel.

[44] DIN Deutsches Institut für Normung e.V. 2014. *Schweißen - Schmelzschweißverbindungen an Stahl, Nickel, Titan und deren Legierungen (ohne Strahlschweißen). Bewertungsgruppen von Unregelmäßigkeiten,* 5817. Accessed 10 October 2020.

[45] DIN Deutsches Institut für Normung e.V. 2018. *Additive Fertigung. Grundlagen_- Terminologie.* Beuth verlag GmbH, Berlin, 52900.

[46] Dinsdale, A. T. and Quested, P. N. 2004. The viscosity of aluminium and its alloys. A review of data and models. *Journal of Materials Science,* 39, 7221–7228.

[47] Dou, L., Yuan, Z. F., Li, J. Q., Li, J., and Wang, X. Q. 2008. Surface tension of molten Al-Si alloy at temperatures ranging from 923 to 1123 K. *Chinese Science Bulletin* 53, 17, 2593–2598.

[48] Dri, N. N., Mindt, H.-W., Shula, B., Megahed, M., Peralta, A., and Neumann, J. DMLS process modelling and validation. In *TMS 2015 144th Annual Meeting & Exhibition*, 389–396.

[49] Eisen, M. A. 2010. *Optimierte Parameterfindung und prozessorientieres Qualitätsmanagement für das Selective Laser Melting Verfahren*. Dissertation. Shaker Verlag, Aachen.

[50] Emmelmann, C. 2019. *Status und Zukunft der Additiven Produktion*. AM Forum Berlin, Berlin.

[51] Engineers Edge. 2020. *Convective Heat Transfer Coefficients Table Chart*. https://www.engineersedge.com/heat_transfer/convective_heat_transfer_coefficients__13378.htm. Accessed 27 March 2020.

[52] Everton, S. K., Hirsch, M., Stravroulakis, P., Leach, R. K., and Clare, A. T. 2016. Review of in-situ process monitoring and in-situ metrology for metal additive manufacturing. *Materials & Design* 95, 431–445.

[53] Fan, Z. 2013. *Numerical modeling of heat transfer and fluid flow in laser metal deposition by powder injection*. Doctoral Dissertations, Missouri University of Science and Technology.

[54] Ferro, P., Meneghello, R., Savio, G., and Berto, F. 2020. A modified volumetric energy density–based approach for porosity assessment in additive manufacturing process design. *Int J Adv Manuf Technol* 110, 7-8, 1911–1921.

[55] Finfrock, C. B., Exil, A., Carroll, J. D., and Deibler, L. 2018. Effect of Hot Isostatic Pressing and Powder Feedstock on Porosity, Microstructure, and Mechanical Properties of Selective Laser Melted AlSi10Mg. *Metallogr. Microstruct. Anal.* 7, 4, 443–456.

[56] Flow Science. 2018. *Flow 3D v11.1 Theory Manual*. Santa Fe, NM: Flow Science, Inc. https://www.flow3d.com

[57] Flow3D. 2017. *Flow3D Reference Manual v11.1*. Santa Fe, NM: Flow Science, Inc. https://www.flow3d.com

[58] Förster, R. and Förster, A. 2018. Einteilung der Fertigungsverfahren nach DIN 8580. *Einführung in die Fertigungstechnik,* Kapitel 2, 23–136.

[59] Francois, M. M., Sun, A., King, W. E., Henson, N. J., Tourret, D., Bronkhorst, C. A., Carlson, N. N., Newman, C. K., Haut, T., Bakosi, J., Gibbs, J. W., Livescu, V., Vander Wiel, S. A., Clarke, A. J., Schraad, M. W., Blacker, T., Lim, H., Rodgers, T., Owen, S., Abdeljawad, F., Madison, J., Anderson, A. T., Fattebert, J.-L., Ferencz, R. M., Hodge, N. E., Khairallah, S. A., and Walton, O. 2017. Modeling of additive manufacturing processes for metals. Challenges and opportunities. *Current Opinion in Solid State and Materials Science* 21, 4, 198–206.

[60] Fraunhofer-Einrichtung für Additive Produktionstechnologien IAPT. 2020.000Z. *Machinery & Tooling - Fraunhofer IAPT*. https://www.iapt.fraunhofer.de/de/Branchenloesungen/Machinery_Tooling.html. Accessed 31 December 2020.609Z.

[61] Gan, Y., Wang, W., Cui, Z., Yan, X., Guan, Z., and Xu, B. 2015. Numerical and experimental study of the temperature field evolution of Mg alloy during high power diode laser surface melting. *Optik - International Journal for Light and Electron Optics* 126, 7-8, 739–743.

[62] Gan, Z., Yu, G., He, X., and Li, S. 2017. Surface-active element transport and its effect on liquid metal flow in laser-assisted additive manufacturing. *International Communications in Heat and Mass Transfer* 86, 206–214.

[63] 2020. *GE Additive eröffnet neues Werk in Deutschland | GE Newsroom*. https://www.genewsroom.com/press-releases/ge-additive-er%C3%B6ffnet-neues-werk-deutschland. Accessed 7 March 2020.

[64] GE Aviation. 2017. *GE Additive unveils first BETA machine from its Project Atlas program*. https://www.ge.com/additive/press-releases/ge-additive-unveils-first-beta-machine-its-project-atlas-program.

[65] Gebhardt, A. 2013. *Generative Fertigungsverfahren. Additive Manufacturing und 3D Drucken für Prototyping - Tooling - Produktion*. Hanser, München.

[66] Ghandi, A., Magar, C., and Roberts, R. 2013. *How Technology can drive the next wave of mass customization*. http://www.mckinsey.com/~/media/mckinsey/dotcom/client_service/bto/pdf/mobt32_02-09_masscustom_r4.ashx. Accessed 21 April 2017.

[67] Goffin, N. J., Higginson, R. L., and Tyrer, J. R. 2015. The use of holographic optical elements (HOE's) to investigate the use of a flat irradiance profile in the control of heat absorption in wire-fed laser cladding. *Journal of Materials Processing Technology* 220, 191–201.

[68] Goffin, N. J., Higginson, R. L., and Tyrer, J. R. 2016. Using wire shaping techniques and holographic optics to optimize deposition characteristics in wire-based laser cladding. *Proceedings. Mathematical, physical, and engineering sciences* 472, 2196, 20160603.

[69] Gong, H., Hengfeng, G., Zeng, K., Dilip, J.J.S., Pal, D., Stucker, B., Christiansen, D., Beuth, J., and Lewandowski, J. J. Melt Pool Characterization for Selective Laser Melting of Ti-6Al-4V Pre-alloyed Powder. In *25th Annual International Solid Freeform Fabrication Symposium*, 256–267.

[70] Grützmann, C. and Wischeropp, T. M. 2014. *A380 Fuel Connector unter Anwendung generativer Fertigungsverfahren*. 13. Fachtagung Rapid Prototyping, HAW Hamburg.

[71] Gu, B., Helvajian, H., Piqué, A., Dunsky, C. M., and Liu, J., Eds. 2017. *Laser 3D Manufacturing IV*. SPIE Proceedings. SPIE.

[72] Gu, D., Hagedorn, Y. C., Meiners, W., Meng, G., Batista, R. J. S., Wissenbach, K., and Poprawe, R. 2012. Densification behavior, microstructure evolution, and wear performance of selective laser melting processed commercially pure titanium. *Acta Materialia* 60, 9, 3849–3860.

[73] Gu, D., Ma, C., Xia, M., Dai, D., and Shi, Q. 2017. A Multiscale Understanding of the Thermodynamic and Kinetic Mechanisms of Laser Additive Manufacturing. *Engineering* 3, 5, 675–684.

[74] Gu, D. and Shen, Y. 2007. Balling phenomena during direct laser sintering of multi-component Cu-based metal powder. *Journal of Alloys and Compounds* 432, 1-2, 163–166.

[75] Gu, D. and Yuan, P. 2015. Thermal evolution behavior and fluid dynamics during laser additive manufacturing of Al-based nanocomposites. Underlying role of reinforcement weight fraction. *Journal of Applied Physics* 118, 23.

[76] Gu, D. D., Meiners, W., Wissenbach, K., and Poprawe, R. 2013. Laser additive manufacturing of metallic components. Materials, processes and mechanisms. *International Materials Reviews* 57, 3, 133–164.

[77] Gunenthiram, V., Peyre, P., Schneider, M., Dal, M., Coste, F., and Fabbro, R. 2017. Analysis of laser–melt pool–powder bed interaction during the selective laser melting of a stainless steel. *Journal of Laser Applications* 29, 2.

[78] Guo, Y., Jia, L., Kong, B., Wang, N., and Zhang, H. 2018. Single track and single layer formation in selective laser melting of niobium solid solution alloy. *Chinese Journal of Aeronautics* 31, 4, 860–866.

[79] Gürtler, F. J., Karg, M., Dobler, M., Kohl, S., Tzivilsky, I., and Schmidt, M. Influence of powder distribution on process stability in laser beam melting: Analysis of melt pool dynamics by numerical simulations. In *25th Annual international solid freeform fabrication symposium*, 1099–1117.

[80] Gürtler, F.-J., Karg, M., Leitz, K.-H., and Schmidt, M. 2013. Simulation of Laser Beam Melting of Steel Powders using the Three-Dimensional Volume of Fluid Method. *Physics Procedia* 41, 881–886.

[81] HALaser Systems. 2017. RepliSLS3D. User manual for 3D SLS/SLM Printer Software, Version 1.7.

[82] Han, L. and Liou, F. W. 2004. Numerical investigation of the influence of laser beam mode on melt pool. *International Journal of Heat and Mass Transfer* 47, 19-20, 4385–4402.

[83] Han, X., Tang, X., Wang, T., Shao, C., Lu, F., and Cui, H. 2018. Role of ambient pressure in keyhole dynamics based on beam transmission path method for laser welding on Al alloy. *Int J Adv Manuf Technol* 99, 5-8, 1639–1651.

[84] Hansen, K. S., Kristiansen, M., and Olsen, F. O. 2014. Beam Shaping to Control of Weldpool Size in Width and Depth. *Physics Procedia* 56, 467–476.

[85] Harvey, J.-P. and Chartrand, P. 2010. Modeling the Hydrogen Solubility in Liquid Aluminum Alloys. *Metallurgical and Materials Transactions B* 41, 4, 908–924.

[86] Higginson, R. L., Gibson, M., Kell, J., and Tyrer, J. 2010. Weld Pool Shaping and Microstructural Control Using Novel Computer Generated Holographic Optic Laser Welding of Steel and Stainless Steel. *Materials Science Forum* 638-642, 3673–3678.

[87] Hirano, K., Fabbro, R., and Muller, M. 2011. Experimental determination of temperature threshold for melt surface deformation during laser interaction on iron at atmospheric pressure. *Journal of Physics D: Applied Physics* 44, 43, 435402.

[88] Hirt, C.W. and Nichols, B.D. 1981. Volume of fluid (VOF) method for the dynamics of free boundaries. *Journal of Computational Physics* 39, 1, 201–225.

[89] Hirt, T. 1998. Thermocapillary Sample Problems, FSI-98-TN47.

[90] Holo'or. *M-Shaper for Scanning Applications. Product Brochure.* https://www.holoor.co.il/wp-content/uploads/2018/03/M-Shaper_brochure.pdf?x67353. Accessed 7 April 2019.

[91] Hovig, E., W., Holm, H., D., and Sorby, K. 2018. Effect of processing parameters on the relative density of AlSi10Mg processed by laser powder bed fusion. *Advanced Manufacturing and Automation VIII.*

[92] Hyer, H., Le Zhou, Park, S., Gottsfritz, G., Benson, G., Tolentino, B., McWilliams, B., Cho, K., and Sohn, Y. 2020. Understanding the Laser Powder Bed Fusion of AlSi10Mg Alloy. *Metallography, Microstructure, and Analysis* 9, 4, 484–502.

[93] Jay, M. L. 2019. *Sustainability: The Future of Manufacturing.* http://compositesmanufacturingmagazine.com/2019/03/sustainability-the-future-of-manufacturing/.

[94] K. S. Hansen, F. O. Olsen, M. Kristiansen, and O. Madsen. 2015. Joining of multiple sheets in a butt-joint configuration using single pass laser welding with multiple spots. *Journal of Laser Applications* 27, 3, 32011.

[95] Kell, J., Tyrer, J. R., Higginson, R. L., Jones, J. C., and Noden, S. 2013. Laser weld pool management through diffractive holographic optics. *Materials Science and Technology* 28, 3, 354–363.

[96] Kempen, K., Thijs, L., van Humbeeck, J., and Kruth, J.-P. 2015. Processing AlSi10Mg by selective laser melting. Parameter optimisation and material characterisation. *Materials Science and Technology* 31, 8, 917–923.

[97] Khairallah, S. A. and Anderson, A. 2014. Mesoscopic simulation model of selective laser melting of stainless steel powder. *Journal of Materials Processing Technology* 214, 11, 2627–2636.

[98] Khairallah, S. A., Anderson, A. T., Rubenchik, A., and King, W. E. 2016. Laser powder-bed fusion additive manufacturing. Physics of complex melt flow and formation mechanisms of pores, spatter, and denudation zones. *Acta Materialia* 108, 36–45.

[99] Khan, M., Jaffrey, S. H. I., Sheikh, N., and Ali, L. 2014. Numerical Simulation of Meltpool Instability in the Selective Laser Melting (SLM) Process. *Lasers in Engineering,* Vol. 28, 319–336.

[100] Kidess, A., Kenjereš, S., Righolt, B. W., and Kleijn, C. R. 2016. Marangoni driven turbulence in high energy surface melting processes. *International Journal of Thermal Sciences* 104, 412–422.

[101] King, W. E., Anderson, A. T., Ferencz, R. M., Hodge, N. E., Kamath, C., Khairallah, S. A., and Rubenchik, A. M. 2015. Laser powder bed fusion additive manufacturing of metals; physics, computational, and materials challenges. *Applied Physics Reviews* 2, 4.

[102] Klahn, C. 2015. *Laseradditiv gefertigte, luftdurchlässige Mesostrukturen. Herstellung und Eigenschaften für die Anwendung*. Springer Berlin Heidelberg, Berlin, Heidelberg.

[103] Klein, C. *Titan-Teil: Bugatti bringt Bremssattel aus dem 3D-Drucker*. https://www.automobil-produktion.de/technik-produktion/produktionstechnik/titan-teil-bugatti-bringt-bremssattel-aus-dem-3d-drucker-120.html. Accessed 23 August 2018.

[104] Koo, B. S. 2013. *Simulation of Melt Penetration and Fluid Flow Behavior during Laser Welding*. Doctor of Philosophy, Lehigh University.

[105] Körner, C., Attar, E., and Heinl, P. 2011. Mesoscopic simulation of selective beam melting processes. *Journal of Materials Processing Technology* 211, 6, 978–987.

[106] Körner, C., Bauereiß, A., and Attar, E. 2013. Fundamental consolidation mechanisms during selective beam melting of powders. *Modelling and Simulation in Materials Science and Engineering* 21, 8.

[107] KPMG. 2014. *Megatrends. Industrial Manufacturing*. Studie. Accessed 21 April 2017.

[108] Kristiansen, M., Hansen, K. S., Langbak, A., Johansen, S. B., Krempin, S. B., and Hornum, M. D. 2017. Single Pass Laser Welding with Multiple Spots to Join Four Sheets in a Butt-joint Configuration. *Physics Procedia* 89, 205–213.

[109] Kruth, J. P., Froyen, L., Rombouts, M., van Vaerenbergh, J., and Mercells, P. 2003. New Ferro Powder for Selective Laser Sintering of Dense Parts. *CIRP Annals - Manufacturing Technology* 52, 1, 139–142.

[110] Kruth, J.-P., Levy, G., Klocke, F., and Childs, T.H.C. 2007. Consolidation phenomena in laser and powder-bed based layered manufacturing. *CIRP Annals - Manufacturing Technology* 56, 2, 730–759.

[111] Kuntz, D. 2012. *Efficient Transformation of Gaussian Beams into Uniform, Rectangular Intensity Distributions*. Accessed 29 December 2017.

[112] Küpper, D., Heising, W., Corman, G., Wolfgang, M., Knizek, C., and Lukic, V. 2017. *Get Ready for Industrialized Additive Manufacturing*. https://www.bcg.com/de-de/publications/2017/lean-manufacturing-industry-4.0-get-ready-for-industrialized-additive-manufacturing.aspx.

[113] Küsters, Y. and Schäfer, M. 2009. Pulverbettbasiertes Strahlschmelzen – neue Forschungserkenntnisse. *RTejournal - Forum für Rapid Technologie* 6.

[114] Lange, F. 2017. *Modellierung der Schmelzbaddynamik beim gepulsten Laserschweißen*. Masterarbeit, Technische Universität Berlin.

[115] Lange, F., Hein, C., and Emmelmann, C. 2018. Numerical optimization of active heat sinks considering restrictions of selective laser melting. *Proceedings of COMSOL Conference Lausanne*.

[116] Laoui, T., Wang, X., Childs, T. H. C., Kruth, J. P., and Froyen, L. 2000. Laser penetration in a powder bed during selective laser sintering of metal powders. Simulations versus experiments. *Proceedings of the Solid Freeform Fabrication*, 453–460.

[117] Laskin, A. Refractive Field Mapping Beam Shaping Optics. Important Feaures for a Right Choice. In *International Congress on Applications 2010*.

[118] Laskin, A. 2012. Creating round and square flattop laser spots in microprocessing systems with scanning optics. *Journal of Laser Applications*.

[119] Lee, Y. 2015. *Simulation of Laser Additive Manufacturing and its Applications. Dissertation*.

[120] Lee, Y. and Zhang, W. Mesoscopic Simulation of Heat Transfer and Fluid Flow in Laser Powder Bed Additive Manufacturing. In *26th Solid Freeform Fabrication Symposium*, 1154–1165.

[121] Lee, Y. S., Nandwana, P., and Zhang, W. 2018. Dynamic simulation of powder packing structure for powder bed additive manufacturing. *The International Journal of Advanced Manufacturing Technology* 96, 1-4, 1507–1520.

[122] Lei, Y. P., Murakawa, H., Shi, Y. W., and Li, X. Y. 2001. Numerical analysis of the competitive infuence of Marangoni fow and evaporation on heat surface temperature and molten pool shape in laser surface remelting. *Computational Material Science* 21, 276–290.

[123] Leitz, K.-H., Grohs, C., Singer, P., Tabernig, B., Plankensteiner, A., Kestler, H., and Sigl, L. S. 2018. Fundamental analysis of the influence of powder characteristics in Selective Laser Melting of molybdenum based on a multi-physical simulation model. *International Journal of Refractory Metals and Hard Materials* 72, 1–8.

[124] Leitz, K.-H., Singer, P., Plankensteiner, A., Tabernig, B., Kestler, H., and Sigl, L. S. 2016. Multi-physical simulation of selective laser melting. *Metal Powder Report* 72, 5, 331–338.

[125] Leitz, K.-H., Singer, P., Plankensteiner, A., Tabernig, B., Kestler, H., and Sigl, L. S. 2016. Thermo-Fluiddynamical Modeling of Laser Beam-Matter Interaction in Selective Laser Melting. *Proceedings of the 2016 COMSOL Conference in Munich*.

[126] Leitz, K.-H., Singer, P., Plankensteiner, A., Tabernig, B., Kestler, H., and Sigl, L. S. 2017. Thermo-fluid Dynamical Simulation of Layer Build-up by Selective Laser Melting of Molybdenum and Steel. *Berg Huettenmaenn Monatsh* 162, 5, 172–178.

[127] Li, C., Fu, C. H., Guo, Y. B., and Fang, F. Z. 2016. A multiscale modeling approach for fast prediction of part distortion in selective laser melting. *Journal of Materials Processing Technology* 229, 703–712.

[128] Li, R., Liu, J., Shi, Y., Wang, L., and Jiang, W. 2012. Balling behavior of stainless steel and nickel powder during selective laser melting process. *The International Journal of Advanced Manufacturing Technology* 59, 9-12, 1025–1035.

[129] Li, W., Zhang, X., and Liou, F. 2018. Modeling analysis of argon gas flow rate's effect on pre-mixed powder separation in laser metal deposition process and experimental validation. *The International Journal of Advanced Manufacturing Technology* 96, 9-12, 4321–4331.

[130] Limmaneevichitr, C. and Kou, S. 2000. Experiments to Simulate Effect of Marangoni Convection on Weld Pool Shape. At high Peclet numbers, the pool bottom can be flat or even convex instead of concave. *WELDING RESEARCH SUPPLEMENT*.

[131] Lin, R., Wang, H.-P., Lu, F., Solomon, J., and Carlson, B. E. 2017. Numerical study of keyhole dynamics and keyhole-induced porosity formation in remote laser welding of Al alloys. *International Journal of Heat and Mass Transfer* 108, Part A, 244–256.

[132] Liu, Y., Yang, Y., Mai, S., Di W., and Song, C. 2015. Investigation into spatter behavior during selective laser melting of AISI 316L stainless steel powder. *Materials & Design* 87, 797–806.

[133] Loh, L. E., Liu, Z. H., Zhang, D. Q., Mapar, M., Sing, S. L., Chua, C. K., and Yeong, W. Y. 2014. Selective Laser Melting of aluminium alloy using a uniform beam profile. *Virtual and Physical Prototyping* 9, 1, 11–16.

[134] Loh, L.-E., Chua, C.-K., Yeong, W.-Y., Song, J., Mapar, M., Sing, S.-L., Liu, Z.-H., and Zhang, D.-Q. 2015. Numerical investigation and an effective modelling on the Selective Laser Melting (SLM) process with aluminium alloy 6061. *International Journal of Heat and Mass Transfer* 80, 288–300.

[135] Loh, L.-E., Song, J., Guo, F., and Bi, G. 2018. Analytical Solution of Temperature Distribution in a Nonuniform Medium Due to a Moving Laser Beam and a Double Beam Scanning Strategy in the Selective Laser Melting Process. *Journal of Heat Transfer,* AUGUST 2018, Vol. 140.

[136] Lu, S., Fujii, H., and Nogi, K. 2004. Marangoni convection and weld shape variations in Ar–O2 and Ar–CO2 shielded GTA welding. *Materials Science and Engineering: A* 380, 1-2, 290–297.

[137] Ly, S., Rubenchik, A. M., Khairallah, S. A., Guss, G., and Matthews, M. J. 2017. Metal vapor micro-jet controls material redistribution in laser powder bed fusion additive manufacturing. *Scientific reports* 7, 1.

[138] Mancisidor, A. M., Garciandia, F., Sebastian, M. S., Álvarez, P., Díaz, J., and Unanue, I. 2016. Reduction of the Residual Porosity in Parts Manufactured by Selective Laser Melting Using Skywriting and High Focus Offset Strategies. *Physics Procedia* 83, 864–873.

[139] Markl, M. and Körner, C. 2016. Multiscale Modeling of Powder Bed–Based Additive Manufacturing. *Annual Review of Materials Research* 46, 93–123.

[140] Markl, M., Rausch, A. M., Küng, V. E., and Körner, C. 2019. SAMPLE: A Software Suite to Predict Consolidation and Microstructure for Powder Bed Fusion Additive Manufacturing. *Adv. Eng. Mater.*, 1901270.

[141] Martin, A. A., Calta, N. P., Khairallah, S. A., Wang, J., Depond, P. J., Fong, A. Y., Thampy, V., Guss, G. M., Kiss, A. M., Stone, K. H., Tassone, C. J., Nelson Weker, J., Toney, M. F., van Buuren, T., and Matthews, M. J. 2019. Dynamics of pore formation during laser powder bed fusion additive manufacturing. *Nature communications* 10, 1, 1987.

[142] Martinez, J.-M., Chesneau, X., and Zeghmati, B. 2006. A new curvature technique calculation for surface tension contribution in PLIC-VOF method. *Computational Mechanics* 37, 2, 182–193.

[143] Matthews, M. J., Guss, G., Khairallah, S. A., Rubenchik, A. M., Depond, P. J., and King, W. E. 2016. Denudation of metal powder layers in laser powder bed fusion processes. *Acta Materialia* 114, 33–42.

[144] Matthews, M. J., Roehling, T. T., Guss, G., Wu, S. Q., Crumb, M. F., Roehling, J. D., and McKeown, J. T. 2018. patial Modulation of Laser Sources for Microstructural Control of Additively Manufactured Metals. *LANE 2018 - 10th Conference on Photonic Technologies,* 00 (2018) 000–000.

[145] MatWeb. 2017. *Aluminium A360.0-F Die Cast Alloy*. http://www.matweb.com/search/GetReference.aspx?matid=9987. Accessed 10 September 2017.

[146] Mazur, M., Benoit, M., Easton, M., and Brandt, M. 2020. Selective laser melting of Inconel 625 alloy with reduced defect formation. *Journal of Laser Applications* 32, 2, 22058.

[147] McFalls, T. A. 2018. *The effect of hydrogen on gas porosity in Laser Powder Bed Fusion of AlSi10Mg*. Master Theses, University of Tennessee.

[148] Megahed, M., Mindt, H.-W., N'Dri, N., Duan, H., and Desmaison, O. 2016. Metal additive-manufacturing process and residual stress modeling. *Integrating Materials and Manufacturing Innovation* 5.

[149] Meiners, W. 1999. *Direktes selektives Laser Sintern einkomponentiger metallischer Werkstoffe*. Berichte aus der Lasertechnik. Shaker, Aachen.

[150] Métais, A., Matteï, S., Tomashchuk, I., Cicala, E., and Gaied, S. 2017. Dissimilar steels laser welding: Experimental and numerical assessment of weld mixing. *Journal of Laser Applications* 29, 2, 22420.

[151] Metal Additive Manufacturing. 2020. *Fraunhofer and Fiat Chrysler additively manufacture sports car wheel components*. https://www.metal-am.com/fraunhofer-and-fiat-chrysler-additively-manufacture-sports-car-wheel-components/. Accessed 22 December 2020.600Z.

[152] 2017. Metal AM. *The Magazine for the Metal Additive Manufacturing Industry,* Vol. 3 No 4.

[153] Metel, A., Stebulyanin, M., Fedorov, S., and Okunkova, A. 2019. Power Density Distribution for Laser Additive Manufacturing (SLM): Potential, Fundamentals and Advanced Applications. *Technologies* 7, 1, 5.

[154] Metelkova, J., Kinds, Y., Kempen, K., Formanoir, C. de, Witvrouw, A., and van Hooreweder, B. 2018. On the influence of laser defocusing in Selective Laser Melting of 316L. *Additive Manufacturing* 23, 161–169.

[155] Midt, H.-W. and Megahed, M. DMLM Models – Numerical Assessment of Porosity. In *22nd International Symposium on Air Breathing Engines*.

[156] Mindt, H. W., Megahed, M., Lavery, N. P., Holmes, M. A., and Brown, S. G. R. 2016. Powder Bed Layer Characteristics. The Overseen First-Order Process Input. *Metallurgical and Materials Transactions A* 47, 8, 3811–3822.

[157] Mindt, H.-W., Desmaison, O., Megahed, M., Peralta, A., and Neumann, J. 2018. Modeling of Powder Bed Manufacturing Defects. *J. of Materi Eng and Perform* 27, 1, 32–43.

[158] Mishra M. 2015. *Shorter Product Life Cycle - An Opportunity or a Challenge?* https://www.linkedin.com/pulse/shorter-product-life-cycle-opportunity-challenge-mahaprasad-mishra. Accessed 21 April 2017.

[159] Mital, A., Desai, A., and Subramanian, A. 2014. *Product Development. A Structured Approach to Consumer Product Development, Design, and Manufacture*. Elsevier Science, Burlington.

[160] Mohanty, S. and Hattel, J. H. 2014. Numerical Model based Reliability Estimation of Selective Laser Melting Process. *Physics Procedia* 56, 379–389.

[161] Möller, M., Vykhtar, B., Emmelmann, C., Li, C. Huang, J., Ed. 2019. *Carbon Footprint and Cost Potential of Additive Manufacturing in Aircraft Systems*.

[162] Moukalled, F., Mangani, L., and Darwish, M. 2016. *The Finite Volume Method in Computational Fluid Dynamics. An Advanced Introduction with OpenFOAM and Matlab*. Fluid Mechanics and Its Applications 113. Springer International Publishing.

[163] MSC Software. 2017. *Simufact Additive. A Simulation Tool for Distortion Prediction in Powder Bed Additive Manufacturing*.

[164] Munsch, M. 2013. *Reduzierung von Eigenspannungen und Verzug in der laseradditiven Fertigung*. Dissertation 2013. Cuvillier Verlag, Göttingen.

[165] Nassar, A. R., Gundermann, M. A., Reutzel, E. W., Guerrier, P., Krane, M. H., and Weldon, M. J. 2019. Formation processes for large ejecta and interactions with melt pool formation in powder bed fusion additive manufacturing. *Scientific reports* 9, 1, 5038.

[166] Okunkova, A., Volosova, M., Peretyagin, P., Vladimirov, Y., Zhirnov, I., and Gusarov, A. V. 2014. Experimental Approbation of Selective Laser Melting of Powders by the Use of Non-Gaussian Power Density Distributions. *Physics Procedia* 56, 48–57.

[167] Okunkova, A., Volosova, M., Peretyagin, P., Zhirnov, I., Podrabinnik, P., Fedorov, S. V., and Gusarov, A. 2016. Study Of Laser Beam Modulation Influence On

Structure Of Materials Produced By Additive Manufacturing. *Advanced Materials Letters* 7, 2, 111–115.

[168] Okunkova, A. A., Peretyagin, P.Y., Podrabinnik, P. A., Zhirnov, I. V., and Gusarov, A. V. 2017. Development of Laser Beam Modulation Assets for the Process Productivity Improvement of Selective Laser Melting. *Procedia IUTAM* 23, 177–186.

[169] Otto, A. and Schmidt, M. 2010. Towards a universal numerical simulation model for laser material processing. *Physics Procedia* 5, 35–46.

[170] Over, C. 2003. *Generative Fertigung von Bauteilen aus Werkzeugstahl X38CrMoV5-1 und Titan TiAl6V4 mit Selective Laser Melting*. Dissertation, RWTH Aachen University.

[171] Pal, D., Patil, N., Zeng, K., and Stucker, B. 2014. An Integrated Approach to Additive Manufacturing Simulations Using Physics Based, Coupled Multiscale Process Modeling. *Journal of Manufacturing Science and Engineering* 136, 6.

[172] Panwisawas, C., Qiu, C., Anderson, M. J., Sovani, Y., Turner, R. P., Attallah, M. M., Brooks, J. W., and Basoalto, H. C. 2017. Mesoscale modelling of selective laser melting: Thermal fluid dynamics and microstructural evolution. *Computational Materials Science* 126, 479–490.

[173] Papadakis, L., Loizou, A., Risse, J., and Schrage, J. 2014. Numerical Computation of Component Shape Distortion Manufactured by Selective Laser Melting. *Procedia CIRP* 18, 90–95.

[174] Parry, L., Ashcroft, I. A., and Wildman, R. D. 2016. Understanding the effect of laser scan strategy on residual stress in selective laser melting through thermo-mechanical simulation. *Additive Manufacturing* 12, Part A, 1–15.

[175] Parteli, E. J. R. DEM simulation of particles of complex shapes using the multi-sphere method. Application for additive manufacturing. In *AIP Conference Proceedings*, 185–188.

[176] Pei, W., Zhengying, W., Zhen, C., Junfeng, L., Shuzhe, Z., and Jun, D. 2017. Numerical simulation and parametric analysis of selective laser melting process of AlSi10Mg powder. *Applied Physics A* 123, 8.

[177] Pothen, M., Winands, K., and Klocke, F. 2017. Compensation of scanner based inertia for laser structuring processes. *Journal of Laser Applications* 29, 1, 12017.

[178] Prashanth, K. G., Scudino, S., Maity, T., Das, J., and Eckert, J. 2017. Is the energy density a reliable parameter for materials synthesis by selective laser melting? *Materials Research Letters* 5, 6, 386–390.

[179] Protasov, C. E. and Gusarov, A. V. 2017. Modeling the Effect of Beam Shaping at Selective Laser Melting. *Procedia IUTAM* 23, 147–154.

[180] Puga Campos, C. A. 2020. *Simulation of the effect of different beam profiles on melt pool size and dynamics during the Selective Laser Melting Process*. Project Thesis, Technical University of Hamburg.

[181] Qiu, C., Panwisawas, C., Ward, M., Basoalto, H. C., Brooks, J. W., and Attallah, M. M. 2015. On the role of melt flow into the surface structure and porosity development during selective laser melting. *Acta Materialia* 96, 72–79.

[182] Rai, R., Palmer, T. A., Elmer, J. W., and Debroy, T. Heat Transfer and Fluid Flow during Electron Beam Welding of 304L Stainless Steel Alloy. *Welding Journal* 2009, 88.

[183] Rausch, A. M., Markl, M., and Körner, C. 2018. Predictive simulation of process windows for powder bed fusion additive manufacturing: Influence of the powder size distribution. *Computers & Mathematics with Applications* 78, 7, 2351–2359.

[184] Rehme, O. 2009. *Cellular Design for Laser Freeform Fabrication*. Dissertation, Technical University of Hamburg.

[185] Rishi Kumar Ganeriwala. *Multiphysics Modeling of Selective Laser Sintering/Melting*. PhD Thesis, University of California, Berkeley.

[186] Roehling, T. T., Wu, S. S. Q., Khairallah, S. A., Roehling, J. D., Soezeri, S. S., Crumb, M. F., and Matthews, M. J. 2017. Modulating laser intensity profile ellipticity for microstructural control during metal additive manufacturing. *Acta Materialia* 128, 197–206.

[187] Rofin. 2016. *Technical Data-Sheet. Coherent FL1000R 1kW*. Accessed 15 August 2019.

[188] RolandBerger, Ed. 2016. *Additive Manufacturing - next generation. Study* 2016.

[189] Rombouts, M. 2006. *Selective Laser Sintering / Melting of Iron-Based Powders*. PhD-Thesis, Katholieke Universiteit Leuven.

[190] Rubenchik, A. M., Khairallah, S., Wu, S., Guss, G., Roehling, T., McKeown, J., and Matthews, M. 2019 - 2019. Beam shape optimization for laser powder beam fusion additive manufacturing (Conference Presentation). In *Laser 3D Manufacturing VI*. SPIE, 10. DOI=10.1117/12.2511202.

[191] Rubenchik, A. M., Wu, S. S., Kanz, V. K., Lowdermilk, W. H., Rotter, M. D., and Stanley, J. 2013. Temperature dependent 780-nm laser absorption by engineering grade aluminum, titanium, and steel alloy surfaces. *Optical Engineering* 53, 12.

[192] Sadrehaghighi, I. 2020. *Mesh Sensitivity in CFD. CFD Open Series*. Patch 1.87.0. https://www.researchgate.net/publication/339289671. Accessed 6 April 2020.

[193] Sahoo, P., Debroy, T., and McNallan, M. J. 1988. Surface tension of binary metal. Surface active solute systems under conditions relevant to welding metallurgy. *Metallurgical Transactions B* 19, 3, 483–491.

[194] Saldi, Z. S. 2012. *Marangoni driven free surface flows in liquid weld pools*. PhD Thesis, Toyohashi University of Technology.

[195] Saldi, Z. S., Kidess, A., Kenjereš, S., Zhao, C., Richardson, I. M., and Kleijn, C. R. 2013. Effect of enhanced heat and mass transport and flow reversal during cool down on weld pool shapes in laser spot welding of steel. *International Journal of Heat and Mass Transfer* 66, 879–888.

[196] Sander, P. 2016. *On the way to Additive Manufacturing*, Hamburg.

[197] ScanLAB. 2011. *varioSCAN 40 Dynamic Focusing Unit. Installation and Operation Handbook*. Rev. 3.2.e.

[198] Scharowsky, T., Bauereiß, A., Singer, R. F., and Körner, C. Observation and numerical simulation of melt pool dynamic and beam powder interaction during selective electron beam melting. In *23rd Annual International Solid Freeform Fabrication Symposium*.

[199] Scheff, S. W., Ed. 2016. *Fundamental statistical principles for the neurobiologist. A survival guide*. Academic Press is an imprint of Elsevier, London.

[200] Schenk, M. 2015. *Produktion und Logistik mit Zukunft*. Springer Berlin Heidelberg, Berlin, Heidelberg.

[201] Schmidt, F. 2015. *3D-Druck hält Einzug im Flugzeugbau | DW | 02.12.2015*. https://www.dw.com/de/3d-druck-h%C3%A4lt-einzug-im-flugzeugbau/a-18886280. Accessed 3 October 2019.

[202] Schmidt, T. 2016. *Potentialbewertung generativer Fertigungsverfahren für Leichtbauteile*. Springer Berlin Heidelberg, Berlin, Heidelberg.

[203] Schoinochoritis, B., Chantzis, D., and Salonitis, K. 2017. Simulation of metallic powder bed additive manufacturing processes with the finite element method. A critical review. *Proceedings of the IMechE* 231, 1, 96–117.

[204] Schwarze, D. 2013. *High Power Selective Laser Melting*. Additive Manufacturing Symposium.

[205] Scipioni Bertoli, U., Wolfer, A. J., Matthews, M. J., Delplanque, J.-P. R., and Schoenung, J. M. 2017. On the limitations of Volumetric Energy Density as a design parameter for Selective Laser Melting. *Materials & Design* 113, 331–340.

[206] Semak, V. and Matsunawa, A. 1997. The role of recoil pressure in energy balance during laser materials processing. *Journal of Physics D: Applied Physics* 30, 18, 2541–2552.

[207] Shi, R., Khairallah, S., Heo, T. W., Rolchigo, M., McKeown, J. T., and Matthews, M. J. 2019. Integrated Simulation Framework for Additively Manufactured Ti-6Al-4V. Melt Pool Dynamics, Microstructure, Solid-State Phase Transformation, and Microelastic Response. *JOM,* Vol. 71, No. 10, 3640–3655.

[208] Shi, R., Khairallah, S. A., Roehling, T. T., Heo, T. W., McKeown, J. T., and Matthews, M. J. 2019. Microstructural control in metal laser powder bed fusion additive manufacturing using laser beam shaping strategy. *Acta Materialia* 184, 284–305.

[209] Shi, X., Ma, S., Liu, C., Chen, C., Wu, Q., Chen, X., and Lu, J. 2016. Performance of High Layer Thickness in Selective Laser Melting of Ti6Al4V. *Materials (Basel, Switzerland)* 9, 12.

[210] Simonelli, M., Tuck, C., Aboulkhair, N. T., Maskery, I., Ashcroft, I., Wildman, R. D., and Hague, R. 2015. A Study on the Laser Spatter and the Oxidation Reactions During Selective Laser Melting of 316L Stainless Steel, Al-Si10-Mg, and Ti-6Al-4V. *Metallurgical and Materials Transactions A* 46, 9, 3842–3851.

[211] SLM Solutions. 2017. *SLM Solutions Group AG: largest single contract in the history of SLM Solutions*. https://slm-solutions.com/investor-relations/announcements/corporate-news/slm-solutions-group-ag-largest-single-contract.

[212] SLM Solutions. 2021. *Peak Productivity: SLM Solutions launches 12-Laser Machine*. https://www.slm-solutions.com/company/news/detail/peak-productivity-slm-solutions-launches-12-laser-machine/. Accessed 2 January 2021.254Z.

[213] SLM Solutions AG. 2020. *SLM Solutions Group AG: BEAMIT unterzeichnet Absichtserklärung für den Kauf von 15 Maschinen von SLM Solutions*. Accessed 7 March 2020.

[214] Song, J., Chew, Y., Jiao, L., Yao, X., Moon, S. K., and Bi, G. 2018. Numerical study of temperature and cooling rate in selective laser melting with functionally graded support structures. *Additive Manufacturing* 24, 543–551.

[215] Soylemez, E. 2020. High deposition rate approach of selective laser melting through defocused single bead experiments and thermal finite element analysis for Ti-6Al-4V. *Additive Manufacturing* 31, 100984.

[216] Steen, W. M. and Mazumder, J. 2010. *Laser Material Processing*. Laser MAterial Processing. Springer, London.

[217] Suchmacher, M. and Geller, M., Eds. 2013. *Practical biostatistics. A user-friendly approach for evidence-based medicine*. Elsevier, Amsterdam, Boston, Heidelberg [etc].

[218] Sufiiarov, V.S., Popovich, A. A., Borisov, E. V., Polozov, I. A., Masaylo, D. V., and Orlov, A. V. 2017. The Effect of Layer Thickness at Selective Laser Melting. *Procedia Engineering* 174, 126–134.

[219] Sundqvist, J. 2015. *Heat Conduction Effects During Laser Welding*. Licentiate Thesis, Luleå University of Technology.

[220] Taheri, S. M. and Hesamian, G. 2013. A generalization of the Wilcoxon signed-rank test and its applications. *Stat Papers* 54, 2, 457–470.

[221] Thanki, Aditi, Goossens, Louca, Mertens, Raya, Probst, Gabriel, Dewulf, Wim, Witvrouw, Ann, Yang, and Shoufeng. 2019. Study of keyhole-porosities in selective laser melting using X-ray computed tomography. *9th Conference on Industrial Computed Tomography*.

[222] The World Bank. 2017. *Industrial Commodities Prices to Surge in 2017. World Bank*, Washington, USA.

[223] Thijs, L., Verhaeghe, F., Craeghs, T., van Humbeeck, J., and Kruth, J.-P. 2010. A study of the microstructural evolution during selective laser melting of TiAl6V4. *Acta Materialia* 58, 9, 3303–3312.

[224] Thorsten Heelinga, Marcel Gerstgrassera, and Konrad Wegenera. 2017. Investigation of selective laser melting spatter characteristics for single- and multi-beam strategies using high speed imaging. *Lasers in Manufacturing Conference 2017*.

[225] Trevisan, F., Calignano, F., Lorusso, M., Pakkanen, J., Aversa, A., Ambrosio, E. P., Lombardi, M., Fino, P., and Manfredi, D. 2017. On the Selective Laser Melting (SLM) of the AlSi10Mg Alloy. Process, Microstructure, and Mechanical Properties. *Materials* 10, 1.

[226] Tseng, W. C. and Aoh, J. N. 2013. Simulation study on laser cladding on pre-placed powder layer with a tailored laser heat source. *Optics & Laser Technology* 48, 141–152.

[227] UPS. 2017. Smarter Operations: The Value Chain's Vital Role in Digital Evolution.

[228] VDI. 2014. Statusreport. Additive Fertigungsverfahren 2014.

[229] VDI-Gesellschaft Produktion und Logistik. 2014. VDI-Richtlinie 3405. Additive Fertigungsverfahren - Grundlagen, Begriffe, Verfahrensbeschreibungen.

[230] Victor, B. M. 2009. *Custom Beam Shaping for High-Power Fibre Laser Welding*. Master Thesis, Ohio state University.

[231] Wang, Q. J. and Chung, Y.-W., Eds. 2013. *Encyclopedia of tribology*. Springer, New York.

[232] Wang, Z., Ummethala, R., Singh, N., Tang, S., Suryanarayana, C., Eckert, J., and Prashanth, K. G. 2020. Selective Laser Melting of Aluminum and Its Alloys. *Materials (Basel, Switzerland)* 13, 20.

[233] Wegener, K., Spierings, A., and Heeling, T. 2016. *Zukunft additiver Fertigung.* http://www.maschinenmarkt.ch/zukunft%C2%ADadditiver%C2%ADferti-gung%C2%ADa%C2%AD557284/. Accessed 16 June 2017.

[234] Wei, P., Wei, Z., Chen, Z., He, Y., and Du, J. 2017. Thermal behavior in single track during selective laser melting of AlSi10Mg powder. *Applied Physics A* 123, 9.

[235] Weingarten, C., Buchbinder, D., Pirch, N., Meiners, W., Wissenbach, K., and Poprawe, R. 2015. Formation and reduction of hydrogen porosity during selective laser melting of AlSi10Mg. *Journal of Materials Processing Technology* 221, 112–120.

[236] 2020.000Z. *What is the role of the viscosity of flux material in welding? - Quora.* https://www.quora.com/What-is-the-role-of-the-viscosity-of-flux-material-in-welding. Accessed 2 July 2020.

[237] Wischeropp, T. M. 2020. *Digitalisierung als Schlüssel für den industriellen 3D Druck.* 3D Printing Week, Mittelstandstag 3D Druck, Lübeck.

[238] Wischeropp, T. M., Emmelmann, C., Brandt, M., and Pateras, A. 2019. Measurement of actual powder layer height and packing density in a single layer in selective laser melting. *Additive Manufacturing* 28, 176–183.

[239] Wischeropp, T. M., Tarhini, H., and Emmelmann, C. 2020. Influence of Laser Beam Profile on the Selective Laser Melting Process of AlSi10Mg. Proceedings of the International Congress of Applications of Lasers & Electro-Optics (ICALEO® 2019). *Journal of Laser Applications,* Volume 32, Issue 2.

[240] Wohlers, T. 2020. *Wohlers Report. 3D printing and additive manufacturing state of the industry.* WOHLERS ASSOCIATES, Fort Collins.

[241] Worth, W. 2018. *Creating a Sustainable Earth Through Sustainable Manufacturing.* https://www.rodongroup.com/blog/importance-sustainable-manufacturing-earth-day.

[242] Wycisk, E. 2017. *Ermüdungseigenschaften der laseradditiv gefertigten Titanlegierung TiAl6V4.* Springer Berlin Heidelberg, Berlin, Heidelberg.

[243] Wycisk, E., Munsch, M., and Schmidt-Lehr M. 2018. Quality in Additive Manufacturing. Best practices for quality management in Additive Manufacturing. *AM-Power Insights,* Vol. 4.

[244] Xia, M., Gu, D., Yu, G., Dai, D., Chen, H., and Shi, Q. 2016. Selective laser melting 3D printing of Ni-based superalloy. Understanding thermodynamic mechanisms. *Science Bulletin* 61, 13, 1013–1022.

[245] Xiang, Y., Zhang, S., Wei, Z., Li, J., Wei, P., Chen, Z., Yang, L., and Jiang, L. 2018. Forming and defect analysis for single track scanning in selective laser melting of Ti6Al4V. *Appl. Phys. A* 124, 10, 155.

[246] Yadroitsev, I., Bertrand, P., and Smurov, I. 2007. Parametric analysis of the selective laser melting process. *Applied Surface Science* 253, 19, 8064–8069.

[247] Yadroitsev, I., Gusarov, A., Yadroitsava, I., and Smurov, I. 2010. Single track formation in selective laser melting of metal powders. *Journal of Materials Processing Technology* 210, 12, 1624–1631.

[248] Yadroitsev, I., Yadroitsava, I., Bertrand, P., and Smurov, I. 2012. Factor analysis of selective laser melting process parameters and geometrical characteristics of synthesized single tracks. *www.emeraldinsight.com/reprints*.

[249] Yakhot, V. and Orszag, S. A. 1986. Renormalization group analysis of turbulence. I. Basic theory. *Journal of Scientific Computation* 1986, Vol. 1 No. 1.

[250] Yakhot, V. and Smith, L. M. 1992. The renormalization group, the ε-expansion and derivation of turbulence models. *Journal of Scientific Computation,* Vol. 7, No. 1.

[251] Yamada, T., Shobu, T., Yamashita, S., Nishimura, A., Muramatsu, T., and Komizo, Y. Visualization technique for quantitative evaluation in laser welding processes. In *In-situ Studies with Photons, Neutrons and Electrons Scattering II*.

[252] Yap, C. Y., Chua, C. K., Dong, Z. L., Liu, Z. H., Zhang, D. Q., Loh, L. E., and Sing, S. L. 2015. Review of selective laser melting. Materials and applications. *Applied Physics Reviews* 2, 4.

[253] Yin, H. and Emi, T. 2003. Marangoni flow at the gas/melt interface of steel. *Metallurgical and Materials Transactions B* 34, 5, 483–493.

[254] Yuan, P. and Gu, D. 2015. Molten pool behaviour and its physical mechanism during selective laser melting of TiC/AlSi10Mg nanocomposites. Simulation and experiments. *Journal of Physics D: Applied Physics* 48, 3.

[255] Zhang, B., Li, Y., and Bai, Q. 2017. Defect Formation Mechanisms in Selective Laser Melting: A Review. *Chinese Journal of Mechanical Engineering* 30, 3, 515–527.

[256] Zhang, B., Liao, H., and Coddet, C. 2012. Effects of processing parameters on properties of selective laser melting Mg-9%Al powder mixture. *Materials & Design* 34, 753–758.

[257] Zhang, J., Liou, F., Seufzer, W., Newkirk, J., Fan, Z., Liu, H., and Sparks, T. E. Probabilistic Simulation of Solidification Microstructure Evolution During Laser-Based Metal Deposition. In *24th Annual International Solid Freeform Fabrication Symposium*, 739–748.

[258] Zhang, J., Song, B., Wei, Q., Bourell, D., and Shi, Y. 2019. A review of selective laser melting of aluminum alloys: Processing, microstructure, property and developing trends. *Journal of Materials Science & Technology* 35, 2, 270–284.

[259] Zhang, T., Li, H., Liu, S., Shen, S., Xie, H., Shi, W., Zhang, G., Shen, B., Chen, L., Xiao, B., and Wei, M. 2019. Evolution of molten pool during selective laser melting of Ti–6Al–4V. *Journal of Physics D: Applied Physics* 52, 5, 55302.

[260] Zhang, X., Li, W., Cui, W., and Liou, F. 2018. Modeling of worn surface geometry for engine blade repair using Laser-aided Direct Metal Deposition process. *Manufacturing Letters* 15, 1–4.

[261] Zhao, C., Fezzaa, K., Cunningham, R. W., Wen, H., Carlo, F. de, Chen, L., Rollett, A. D., and Sun, T. 2017. Real-time monitoring of laser powder bed fusion process using high-speed X-ray imaging and diffraction. *Scientific reports* 7.

[262] Zhao, C. X., Kwakernaak, C., Pan, Y., Richardson, I. M., Saldi, Z., Kenjeres, S., and Kleijn, C. R. 2010. The effect of oxygen on transitional Marangoni flow in laser spot welding. *Acta Materialia* 58, 19, 6345–6357.

[263] Zhao, L., Tsukamoto, S., Arakane, G., Sugino, T., and Debroy, T. 2013. Influence of oxygen on weld geometry in fibre laser and fibre laser–GMA hybrid welding. *Science and Technology of Welding and Joining* 16, 2, 166–173.

[264] Zhirnov, I., Kotoban, D. V., and Gusarov, A. V. 2018. Evaporation-induced gas-phase flows at selective laser melting. *Applied Physics A* 124, 2, 41101.

[265] Zhirnov, I. V., Podrabinnik, P. A., Okunkova, A. A., Gusarov, A. V., Grigoriev, S. N., and Volosova, M. A. 2015. Laser beam profiling. Experimental study of its influence on single-track formation by selective laser melting. *Mechanics & Industry* 16, 7.

[266] Zhou, J., Zhang, Y., and Chen, J. K. 2009. Numerical simulation of laser irradiation to a randomly packed bimodal powder bed. *International Journal of Heat and Mass Transfer* 52, 13-14, 3137–3146.

[267] Zielinski, J., Düchting, J., Meiners, W., and Schleifenbaum, J. H. 2018. Measurement of the Effective Absorption of Laser Radiation for Different States Relevant in the LB-PBF Process on the Example of Ti6Al4V. *Proceedings of DDMC 2018*.

[268] Zielinski, J., Vervoort, S., Mindt, H.-W., and Megahed, M. 2017. Influence of Powder Bed Characteristics on Material Quality in Additive Manufacturing. *Berg Huettenmaenn Monatsh* 162, 5, 192–198.

[269] Zohdi, T. I. 2013. Rapid Simulation of Laser Processing of Discrete Particulate Materials. *Arch Computat Methods Eng* 20, 4, 309–325.

In the course of this thesis, the following student theses have been created:

Tarhini, H. 2018: Effect of Laser Beam Profile on the Selective Laser Melting Process of Aluminum Alloys. *Master Thesis*, TU Hamburg.

Puga Campos, C. A. 2020: Simulation of the effect of different beam profiles on melt pool size and dynamics during the Selective Laser Melting Process. *Project Thesis*, TU Hamburg.

List of Figures

T. M. Wischeropp, *Advancement of Selective Laser Melting by Laser Beam Shaping*, Light Engineering für die Praxis, https://doi.org/10.1007/978-3-662-64585-7

List of Tables

T. M. Wischeropp, *Advancement of Selective Laser Melting by Laser Beam Shaping*, Light Engineering für die Praxis, https://doi.org/10.1007/978-3-662-64585-7

A. Appendix

A.1 Description of Numerical Model

In the following, a detailed description of the numerical model used in the thesis is provided (see Chapters 3 and 5.1).

For the simulation, the commercial software Flow-3D (V11.2.6.4) from Flow Science Inc. is used. In the software, the domain is divided into a mesh of rectangular cells. Within each cell, the values of the dependent variables are averaged, except for the velocity, which is determined for each cell face [57]. The software distinguishes between *solid*, *fluid* and *void* regions. A *solid* defines a part of the domain with material that cannot change its state (is always solid) but can contribute to the heat conduction. *Fluid* on the other hand defines a material that can change its state from solid to liquid or gas and can flow within the domain when it is in a liquid state. A *void* region describes the part of the domain which is not occupied by a solid or liquid and, therefore, represents the gas region. The gas region is not simulated directly to increase the numerical efficiency, but its effects on the fluid are incorporated by boundary conditions.

The amount of fluid within each cell is described by the fluid fraction coefficient F, which varies between $0 \leq F \leq 1$, as shown in Figure A-1 [139]. A value of $F = 0$ corresponds to a cell containing no fluid and $F = 1$ to a cell completely filled with fluid [53]. For partially filled cells, the coefficient has a value corresponding to its fill fraction, as shown in Figure A-1 [88, 194]. The physical properties (e.g., conductivity, density, …) of partially filled cells are averaged according to the cell's fluid fraction coefficient.

Flow-3D uses the second-order FAVOR^{TM} (*Fractional Area-Volume Obstacle Representation*) algorithm to calculate the surface and its curvature of a cell, based on its fluid fraction as well as the fluid fraction of the adjacent cells for every time-step [53]. The algorithm is similar to the PLIC (Piecewise Linear Interface Construction) method [142] with additional non-disclosed routines that minimize the smearing of the liquid-gas interfaces, a typical side effect of the VOF method [56, 194].

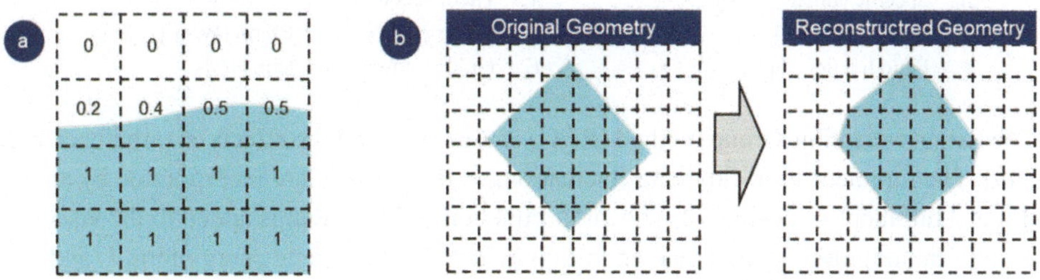

Figure A-1: a) Example of fluid fraction value F in VOF method; b) Example of Surface Reconstruction by FAVOR^{TM} method. (systematic sketches – numbers/graphics are not to scale) (based on [57]).

T. M. Wischeropp, *Advancement of Selective Laser Melting by Laser Beam Shaping*, Light Engineering für die Praxis, https://doi.org/10.1007/978-3-662-64585-7

A.1.1 Governing Equations for Heat and Mass Transfer

For the calculation of the thermal fluid flow, the second order Navier–Stokes equations as well as the VOF equation are solved for each time-step [56, 181].

The Navier-Stokes equations are a set of three non-linear partial differential equations (PDEs), namely the impulse, the mass and the energy conservation equation [48, 181]. They are shown in Equation A.1-A.3 for an incompressible, Newtonian fluid. A specific momentum source vector is added to the impulse equation to implement the effects of surface forces. Furthermore, an energy source term is added to the energy conservation equation to account for the energy input by the laser beam as well as energy loss by vaporization, radiation and convection [8]. The temporal change of the fluid fraction F within each cell is calculated by the Volume of Fluid equation shown in Equation A.4 [56].

Impulse equation

$$\rho\left[\frac{\partial\vec{v}}{\partial t}+(\nabla\cdot\vec{v})\vec{v}\right]=-\nabla p+\mu\nabla^2\vec{v}+\rho\vec{g}+\vec{f} \tag{A.1}$$

Mass conservation equation

$$\frac{\partial\rho}{\partial t}+\nabla\cdot\rho\vec{v}=\mathrm{m_S} \tag{A.2}$$

Energy conservation equation

$$\rho\frac{\partial h}{\partial t}+\nabla\cdot\rho h\vec{v}=-p\nabla\cdot\vec{v}+k\nabla^2T+h_E \tag{A.3}$$

Volume-of-Fluid equation

$$\frac{\partial F}{\partial t}+\nabla\cdot F\vec{v}=\mathrm{F_S} \tag{A.4}$$

ρ: Density of material [kg/m^3]
∇: Nabla operator [1/m]
m_S: Specific mass source term [kg/m^{2}s]
μ: Dynamic viscosity of fluid [kg/(m·s)]
h: Specific enthalpy [J/kg]
T: Temperature [K]
F: Fluid fraction [1]
$\vec{v}$: Velocity vector [m/s]
p: Pressure [kg/(m·s^2)]
$\vec{g}$: Gravity vector [m/s^2]
$\vec{f}$: Specific momentum source vector [kg/(m^2·s^2)]
k: Thermal conductivity [W/(m·K)]
h_E: Specific energy source term [W/m^2]
F_S: Fluid fraction source term [1/s]

The Renormalization-Group method (RNG) is used to model the effects of turbulence [57, 249]. The turbulent fluid flow and fluctuations in the velocity are incorporated by an additional turbulent viscosity v_t as shown in Equation A.5 [57] that is added to the viscosity term of the impulse equation shown in Equation A.1. The kinetic energy term k_t and turbulent energy dissipation ε_t are calculated by two partial differential equations that are described in detail by Yakhot and Smith [57, 250].

$$\mu = \rho(\nu + \nu_t) \tag{A.5}$$

$$\nu_t = 0.085\frac{k_t^2}{\varepsilon_t} \tag{A.6}$$

μ: Dynamic viscosity of fluid [kg/(m·s)]
ρ: Density of material [kg/m³]
ν: Kinematic viscosity of fluid [m²/s]
ν_t: Turbulent kinematic viscosity [m²/s]
k_t: Turbulent kinetic energy [kg²·m²/s³]
ε_t: Turbulent energy dissipation [kg²·m²/s³]

A.1.2 Free Surface Heat and Mass Transfer

In this chapter, an overview of the implemented phenomena for heat and mass transfer at the surface is provided. A more detailed description of the energy input by laser beam, vaporization and solidification is given in the Chapters A.1.3 - A.1.5.

Mass Transfer

Three different forces act on the surface of the melt pool as shown in Figure A-2 and Equation A.7; namely the capillary force $\vec{f_c}$, thermo-capillary force $\vec{f_{tc}}$ and vaporization pressure force $\vec{f_p}$ [11, 97, 139]. The vapor shear stress is ignored due to a minor effect on the SLM process (see Chapter 4.1).

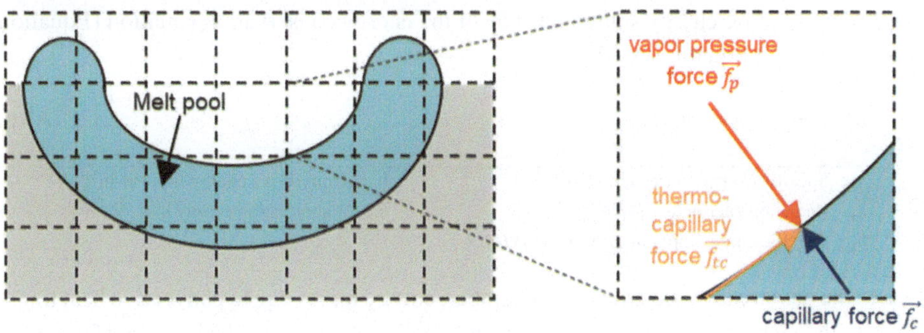

Figure A-2: Forces acting on the surface of the melt pool.

$$\vec{f} = \vec{f_c} + \vec{f_{tc}} + \vec{f_p} \tag{A.7}$$

$\vec{f}$: Specific momentum source vector [kg/(m²·s²)]
$\vec{f_c}$: Capillary force vector [kg/(m²·s²)]
$\vec{f_{tc}}$: Thermo-capillary force vector [kg/(m²·s²)]
$\vec{f_p}$: Vapor pressure force vector [kg/(m²·s²)]

The capillary force $\vec{f_c}$ acts normally to the liquid surface and has a magnitude proportional to the local radius of curvature [89]. The force is calculated by Equation A.8 [114].

$$\vec{f_c} = \sigma(\nabla s \cdot \vec{n})\vec{n} \tag{A.8}$$

$\vec{f_c}$:	Capillary force vector [kg/(m²·s²)]	σ:	Surface tension [N/m²]
s:	Surface curvature [1]	∇:	Nabla operator [1/m]
$\vec{n}$:	Normal vector [1]		

The thermo-capillary force $\vec{f}_{tc}$ acts tangentially to the liquid surface and is proportional to the temperature gradient as well as the dependency of surface tension on temperature [22, 98, 114]. It can be calculated by Equation A.9 [114].

$$\vec{f}_{tc} = \frac{\partial \sigma}{\partial T} \nabla \vec{T} \tag{A.9}$$

$\vec{f_{tc}}$:	Thermo-capillary force vector [kg/(m²·s²)]	σ:	Surface tension [N/m²]
∇:	Nabla operator [1/m]	$\vec{T}$:	Temperature vector [K]

The vaporization pressure induces a force $\vec{f}_p$ normal to the liquid surface. It is described in more detail in Chapter A.1.4.

Mass Transfer

Four phenomena play a crucial role in the energy transport at the free surface, namely convection, radiation and vaporization as well as energy input of the laser beam. The corresponding specific energy source term h_E of the energy conservation equation (Equation A.3) is shown in Equation A.10.

$$h_E = -h_c - h_r - h_v + h_{lb} \tag{A.10}$$

h_E:	Specific energy source term [W/m²]	h_C:	Convective energy loss [W/m²]
h_r:	Radiation energy loss [W/m²]	h_v:	Vaporization energy loss [W/m²]
h_{lb}:	Specific energy input of laser beam [W/m²]		

Due to the gas stream inside the build chamber, energy losses due to convective heat transfer h_c to the ambient gas occur [125]. The energy loss depends on the heat transfer coefficient h_{cv} which is relative to the velocity between the fluid surface and the ambient gas. For the numerical model, it is assumed that the heat transfer coefficient h_{ce}, as well as the void (gas) temperature T_{vo}, are constant. They are implemented in the numerical model as shown in Equation A.11 [56].

$$h_c = h_{ce}(T_s - T_{vo}) \tag{A.11}$$

h_c:	Specific convective energy loss [W/m²]	h_{ce}	Convective heat transfer coefficient [W/m²K]
T_{vo}:	Temperature of the void [K]	T_s:	Temperature of the surface [K]

Additional energy loss at the free surface occurs due to radiation. It can be described by Stefan-Boltzmann law and calculated by the equation shown in A.12 [11, 98].

$$h_r = \sigma_{SB} \cdot \varepsilon(T_s^4 - T_{vo}^4) \tag{A.12}$$

h_r:	Radiation energy loss [W/m²]	σ_{SB}	Stefan-Boltzmann constant [W/m²K⁴]
ε:	Emissivity of the surface [1]	T_s:	Temperature of the surface [K]
T_{vo}:	Temperature of the void [K]		

The energy input by the laser beam h_{lb} is described in detail in Chapter A.1.3 and the energy loss due to vaporization h_v in Chapter A.1.4.

A.1.3 Laser Beam and Multiple-Reflection

To simulate the energy input of the laser beam h_{lb}, a Fresnel-based absorption and multiple-reflections are implemented.

The Fresnel absorption describes the reflection and transmission of an electromagnetic wave on a flat surface and is used in many numerical models to describe the laser beam absorption in laser material processing [80, 131, 219]. The local absorptivity and reflection depend on the emissivity of the surface as well as the inclination angle of the electromagnetic wave as shown in Equation A.13 as well as Figure A-3 [32].

$$a_{fres} = 1 - \frac{1}{2}\left(\frac{1 + (1 - \varepsilon cos\phi)^2}{1 + (1 + \varepsilon cos\phi)^2} + \frac{\varepsilon^2 - 2\varepsilon cos\phi + 2cos^2\phi}{\varepsilon^2 + 2\varepsilon cos\phi + 2cos^2\phi}\right) \tag{A.13}$$

a_{fres}:	Fresnel absorption coefficient [1]	ε:	Emissivity/absorptivity of the surface [1]
ϕ:	Surface-irradiation angle of laser beam [rad; °]		

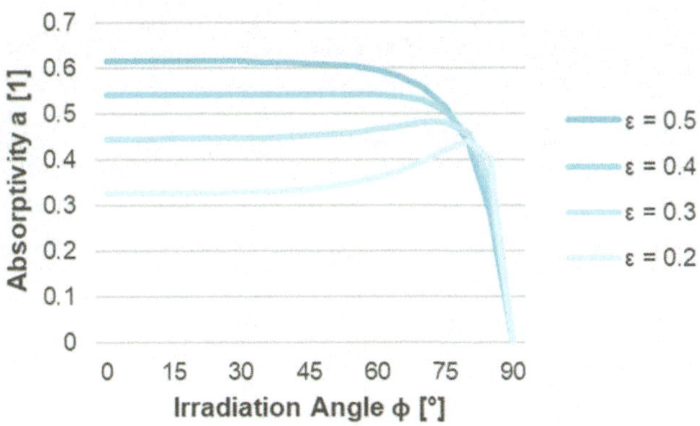

Figure A-3: Fresnel absorption for different emissivities ε and angles of irradiation ϕ.

A part of the reflected electromagnetic wave will interact with the surface of the material several times due to multiple-reflection. A ray-tracing model is implemented in Flow3D to simulate this [98, 203]. Therefore, the laser beam is divided into single rays, each with a specific heat flux h_b corresponding to the local intensity of the laser beam. For each time-step, the overall energy input of the laser beam h_{lb} is calculated by a superposition of the energy input of each laser beam ray (see Equation A.14) and incorporated in Equation A.10.

$$h_{lb} = \sum h_b \cdot a_{fres\_i} \tag{A.14}$$

h_{lb}: Specific energy input of laser beam [W/m²]
h_b: Specific energy input of single laser beam ray [W/m²]
a_{fres_i}: Local absorptivity [1]

Two different laser beams, one with a Gaussian and one with a donut beam profile, are implemented in the numerical model. The formula for the mathematical description of the Gaussian beam profile is shown in Equation A.15. The donut beam profile is created by subtracting a second, smaller Gaussian beam from the first one as shown in Equation A.16.

$$h_{lb\_G}(\mathrm{r}) = \frac{P_L}{\pi \cdot r_b^2} e^{\left(-\frac{2r^2}{r_b^2}\right)} \tag{A.15}$$

$$h_{lb\_D}(\mathrm{r}) = \frac{P_L}{\pi \cdot r_b^2} e^{\left(-\frac{2r^2}{r_b^2}\right)} - \frac{P_{L2}}{\pi \cdot r_{b2}^2} e^{\left(-\frac{2r^2}{r_{b2}^2}\right)} \tag{A.16}$$

h_{lb_G}: Specific energy input of Gaussian laser beam [W/m²]
h_{lb_D}: Specific energy input of donut laser beam [W/m²]
r: Distance from beam center [µm]
r_{b2}: Radius of subtracted laser beam [µm]
P_L: Laser power [W]
P_{L2}: Subtracted laser power [W]
r_b: Radius of laser beam [µm]

Figure A-4 visualizes the implemented Gaussian and donut beam profiles. The parameters are chosen in a way that the resulting beam profiles match the measured beam profiles shown in Figure 4-3 on page 56. Therefore, the following parameters have been defined: $r_b = 140$ µm, $r_{b2} = 70$ µm, $P_{L2} = 0.6 \cdot P_L$.

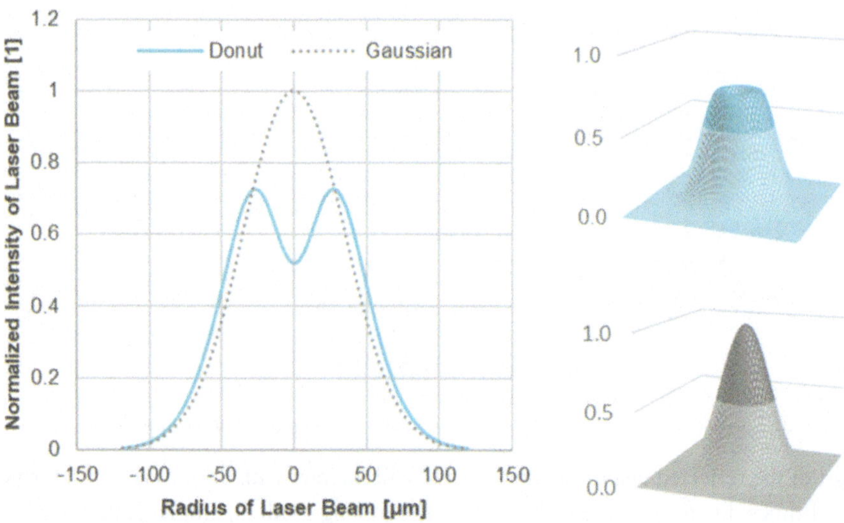

Figure A-4: Visualization of the Gaussian (TEM_{00}) and donut beam profiles implemented in the numerical model.

A.1.4 Vaporization

As described in Chapter 2.2.5, vaporization leads to heat and mass losses and induces pressure on the surface. All these effects are implemented in the numerical model.

When the temperature of the fluid at a free surface exceeds the saturation temperature, the fluid starts to evaporate. The amount of energy h_v that evaporates is proportional to the temperature of the surface and material-specific properties as shown in Equation A.17 [56].

$$h_v = \alpha k_f (T_s - T_b)/l_c \tag{A.17}$$

h_v:	Vaporization energy loss [W/m²]	α:	Accommodation coefficient [1]
k_f:	Thermal conductivity of the fluid [W/(m·K)]	T_s:	Temperature of the surface [K]
T_b:	Boiling temperature of fluid [K]	l_c:	Characteristic length for heat cond. [mm]

The characteristic length l_c for heat conduction in the liquid at the free surface is defined in terms of the Prandtl number as shown in Equation A.18 [56].

$$l_c = x_{min}\sqrt{0.0628 \cdot k_f / C_v \mu} \tag{A.18}$$

l_c:	Characteristic length for heat conduction [mm]	x_{min}:	Minimum cell size [mm]
k_f:	Thermal conductivity of the fluid [W/(m·K)]	μ:	Viscosity of the fluid [kg/(m·s)]
C_v:	Vapor specific heat at constant volume [J/(kg·K)]		

Additionally, the vaporization pressure induces a force $\vec{f_p}$ that is normal to the liquid surface as shown in Figure A-2. Its value depends on the recondensation rate as well as saturation pressure as shown in Equation A.19 and A.20 [56, 143, 206].

$$\vec{f_p} = p_r \vec{n} \tag{A.19}$$

$$p_r = \frac{1 - \beta_R}{2} p_{sat} \tag{A.20}$$

$\vec{f_p}$:	Vaporization pressure force vector [kg/(m²·s²)]	p_r:	Specific vapor pressure [Pa/m]
p_{sat}:	Specific saturation pressure [Pa/m]	β_R:	Recondensation rate [1]
$\vec{n}$:	Normal vector [1]		

The saturation pressure is calculated by the Clausius-Clapeyron-equation shown in Equation A.21 [57, 101, 104]. It depends on material-specific constants, the ambient gas pressure as well as the surface temperature.

$$p_{sat} = p_0 e^{\left[K_1\left(1-\frac{T_b}{T_s}\right)\right]} \tag{A.21}$$

$$K_1 = \frac{\Delta H_v}{R \cdot T_b} \tag{A.22}$$

p_{sat}:	Specific saturation pressure [Pa/m]	p_0:	Specific ambient gas pressure [Pa/m]
K_1:	Vaporization pressure constant [1]	T_b:	Boiling temperature [K]
T_s:	Temperature of the surface [K]	ΔH_v:	Latent heat of vaporization [J/kg]
R:	Gas constant [kg·m²/(s²mol·K)]		

The mass loss due to vaporization is calculated as shown in Equation A.23 [57, 131]. It is implemented in the mass conservation Equation A.2 as a negative mass source term. The corresponding fluid fraction source term F_S of the Volume of Fluid equation (Equation A.4) is calculated by dividing the mass loss term m_{vap} by the temperature-dependent density of the material.

$$m_{vap} = \alpha \sqrt{\frac{M}{2\pi R T_s}} (p_{sat} - p_r) \tag{A.23}$$

m_{vap}:	Specific mass loss due to vaporization [Pa/m]	α:	Accommodation coefficient [1]
M:	Molecular weight of vapor [kg/mol]	R:	Gas constant [kg·m²/(s²mol·K)]
T_s:	Temperature of the surface [K]	p_{sat}:	Specific saturation pressure [kg/(m²·s²)]
p_r:	Specific vapor pressure [kg/(m²·s²)]		

A.1.5 Solidification

During the phase transition from liquid to solid (and vice versa) the latent heat of fusion is released (or absorbed) and the free flow of the material is hindered since the material is partly in liquid and partly in the solid phase [18, 234]. It is assumed that the latent heat of fusion is released linearly and that the solid fraction S decreases linearly with a decrease in temperature between the liquidus temperature T_{LT} and solidus temperature T_{ST}. To incorporate the flow-resistance effect of the solid phase, three different stages are considered, depending on the solid fraction S as shown in Figure A-5 [18, 57, 176].

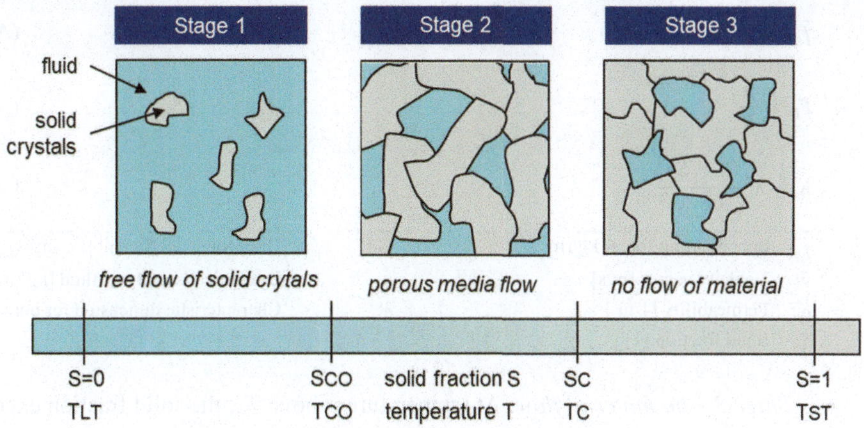

Figure A-5: Different stages during solidification of the melt pool.

- *Stage 1 – free flow of solid crystals*: This stage is defined by the temperature-interval between the liquidus temperature T_{LT} and the temperature T_{CO} at which the solid fraction equals the coherent solid fraction S_{CO} [56]. Within this temperature-interval, the solid crystals can flow freely (i.e., the solid phase is not coherent). It is assumed that the viscosity of the fluid increases linearly with decreasing temperature as defined in Equation A.24 in this stage. Above temperatures of T_{CO}, the viscosity is considered to be constant ($\mu_S = \mu_S(T_{CO})$).

$$\mu_S = \mu_{TL} + k_\mu \cdot (T_{LT} - T) \tag{A.24}$$

μ_S: Dynamic viscosity of fluid during solidification [kg/(m·s)]
μ_{TL}: Dynamic viscosity of fluid at liquidus temperature [kg/(m·s)]
k_μ: Temperature-dependent viscosity gradient [kg/(K·m·s)]
T_{LT}: Liquidus temperature [K]
T: Temperature of the fluid [K]

- *Stage 2 – porous media flow*: This stage is defined by the temperature-interval between T_{CO} and temperature T_C at which the solid fraction equals the critical solid fraction S_C [56]. The solid fraction S_C defines the point at which the solid phase changes from a porous to a rigid structure. Within this temperature interval, the material is treated as a porous medium. A darcy-type drag force $\vec{f_d}$ (see Equation A.25) that is proportional to the fluid velocity $\vec{v}$ is added to the force vector $\vec{f}$ of the impulse equation (Equation A.1) [18]. The drag coefficient F_D of this equation depends on the permeability K of the porous structure as well as the dynamic viscosity μ of the fluid as shown in Equations A.26 and A.27 [18, 56].

$$\overrightarrow{f_d} = F_D \cdot \vec{v} \tag{A.25}$$

$$F_D = \frac{\mu}{K} \tag{A.26}$$

$$K = \lambda^2 \cdot \frac{(1-S)^3}{180 \cdot S^2} \tag{A.27}$$

$\overrightarrow{f_d}$:	Specific drag force [kg/(m²·s²)]	F_D:	Drag force coefficient [kg/(m³·s)]
$\vec{v}$:	Velocity vector [m/s]	μ:	Dynamic viscosity of fluid [kg/(m·s)]
K:	Permeability [1/m²]	λ:	Characteristic dimension for porous media [m]
S:	Solid fraction [1]		

- *Stage 3 – no material flow*: At temperatures above T_C, the solid fraction exceeds the critical solid fraction S_C that defines the point of rigidity. At this stage, the solid fraction forms a closed structure and prevents any material flow. Therefore, the drag force $\overrightarrow{f_d}$ of Equation A.25 is defined as infinite [18, 56].

A.2 Mesh Convergence Study

A fine mesh is necessary to simulate the melt pool dynamics in the required detail. However, since the computational time increases cubically with a reduction in mesh size, a trade-off between accuracy and numerical efficiency must be made [192]. To find the optimum balance between these requirements, a mesh convergence study is done. In this regard, the melting of single tracks with different mesh sizes is simulated and key characteristics of the results are compared. The parameters used for the simulation are listed in Table A-1 along with the relevant analyzed characteristics. The results of the simulations are shown in Figure A-6 and Figure A-7.

Table A-1: Parameters used for mesh convergence study.

Parameter	Value	Characteristics Analyzed
Laser power P_L [W]	700 W	• Melt pool dimensions • Melt-velocities • Melt pool temperatures • Calculation time
Scanning speed v_S [mm/s]	2,500 mm/s	
Temp. build-platform T_{BP}	20 °C	
Laser beam profile	Gaussian (G)	
Laser beam diameter d_f	140 µm	
Mesh size	10, 15, 20, 25, 30 µm	

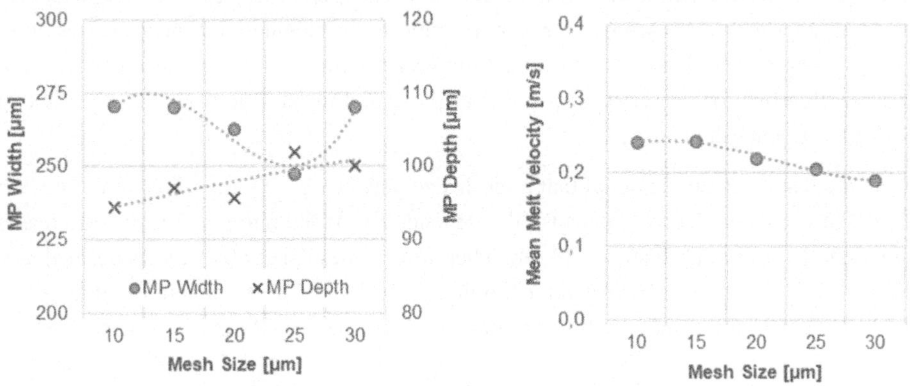

Figure A-6: Melt pool (MP) dimensions and mean melt velocities depending on mesh size.

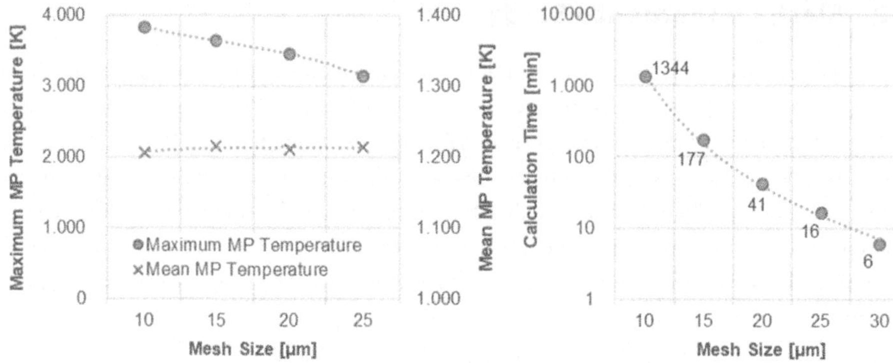

Figure A-7: Melt pool (MP) temperatures and calculation time, depending on mesh size.

From the results, it can be seen that there is a relevant dependency of the melt pool characteristics on the mesh size. For the melt pool dimensions and melt velocities, the effect of the mesh size is quite low between 10-20 µm but gets more significant for coarser meshes. The mean melt pool temperature only shows a very little effect on the mesh size, while the maximum melt pool temperature is linearly decreasing with an increase in mesh size. The fact that the maximum melt pool temperature increases with a reduction in mesh size is an expected result as the finer resolution in the area of highest laser intensity will lead to higher maximum temperatures. Furthermore, the calculation time increases cubically. It must be said that no stable simulation was possible with a mesh size of 7.5 µm or smaller. Reasons for this could numerical artifacts and instabilities that can come along with a too fine mesh [13, 162].

From the results, it can be noted that a mesh size between 10-20 µm seems to be the best choice in terms of accuracy, because the dependency of the numerical results increases significantly with a mesh size > 20 µm. Therefore, a mesh size of 15 µm is defined as a mesh size of 10 µm leads to very long simulations times and as the resolution of the powder bed is conceived as not accurate enough at a mesh size of 20 µm (see Figure A-8).

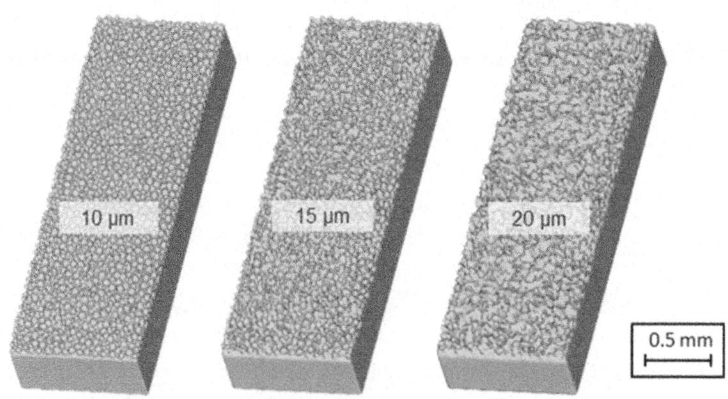

Figure A-8: Visualization of powder bed for mesh sizes of 10, 15 and 20 µm.

A.3 Image-Processing for Spatter Analysis

In the following, the processing of the images for the spatter analysis is described (see Chapters 4.4 and 5.3.1). The processing is done with a MATLAB script that was developed together with Hadi Tarhini during his internship at the Fraunhofer IAPT in 2019.

The MATLAB script converts every single image of the high-speed video into binary images with *bright* and *dark* spatters (see Figure 4-7). *Bright* spatters are defined as spatters that have a brightness above a defined threshold, and *dark* spatters below this threshold. It is assumed that there is a high correlation between bright spatters and spatters that are ejected directly from the melt pool as well as dark spatters and spatters that are the result of entrained powder particles. For the processing, the following processing steps are performed:

- The image is converted into a greyscale picture that only contains information about the brightness of the pixels. The brightness has a value between 0 and 255.
- The noise of the frame is reduced by subtracting the average brightness of each pixel over the whole video from the specific frame that is analyzed.
- The grey-scale image is converted into a binary image to count the bright spatters. A threshold brightness of 90 is chosen for the conversion. This threshold was experimentally determined and defined as the size of the spatters of the resulting binary image matches the size of the *bright* spatters of the original image.
    - all bright spatters are identified by the *regionprops* function of MATLAB. The function counts the number of separated regions of the binary image with connected pixels of a value of "1".
    - the melt pool is removed from the binary image by deleting the largest identified region of the image.
    - the amount of spatters is counted by the *regionprops* function.
- The greyscale image is converted into a binary image to count the dark spatters. A threshold at a brightness of 20 is chosen for the conversion. This threshold was experimentally determined and defined as the size of the spatters of the resulting binary image matches the size of the *dark* spatters of the original image.
    - all bright and dark spatters are identified by the *regionprops* function of MATLAB.
    - the melt pool is removed from the binary image by deleting the largest identified region of the image.

- the *bright* spatters are removed from the binary image by deleting all regions that have been identified in the binary image of the *bright* spatters.
- the amount of spatters is counted by the *regionprops* function.

With the help of this MATLAB routine, the amount of dark and bright spatters of every single frame of a video is counted and the average amount of spatters, as well as its standard deviation, are calculated over 200 frames.

A.4 Results of Spatter Analysis

Table A-2 summarizes the measured amount of spatters during the experiments described in Chapter 5.3.1.

Table A-2: Average amount and standard deviation of spatters measured during the multi-layer SLM process for the Gaussian and donut beam profiles.

Laser Power [W]	Scan Speed [mm/s]	Hatch-Distance [mm]	Layer thickness [mm]	VED [J/mm³]	Build Rate [mm³/s]	Amount of Spatters for Gaussian Beam Profile		Amount of Spatters for Donut Beam Profile	
						AVG	STD	AVG	STD
400	1,000	0.1	0.05	80.0	5.0	15.0	4.8	8.9	4.3
400	1,500	0.1	0.05	53.3	7.5	14.8	4.4	7.4	3.6
400	2,000	0.1	0.05	40.0	10.0	10.3	5.2	7.5	3.6
600	1,000	0.1	0.05	120.0	5.0	17.6	5.9	6.9	3.5
600	1,500	0.1	0.05	80.0	7.5	21.0	8.3	11.4	6.6
600	2,000	0.1	0.05	60.0	10.0	18.4	6.7	9.0	4.8
800	1,000	0.1	0.05	160.0	5.0	10.2	5.7	5.5	6.2
800	1,500	0.1	0.05	106.7	7.5	12.8	6.1	5.4	5.1
800	2,000	0.1	0.05	80.0	10.0	10.8	6.4	6.0	4.3
400	1,000	0.15	0.05	53.3	7.5	11.9	6.6	8.7	4.8
400	1,500	0.15	0.05	35.6	11.3	13.6	5.9	8.4	4.6
400	2,000	0.15	0.05	26.7	15.0	13.0	6.1	7.8	4.4
600	1,000	0.15	0.05	80.0	7.5	11.7	5.0	6.9	5.8
600	1,500	0.15	0.05	53.3	11.3	13.0	5.7	8.4	6.4
600	2,000	0.15	0.05	40.0	15.0	12.8	5.6	8.2	6.0
800	1,000	0.15	0.05	106.7	7.5	12.3	5.7	6.8	5.8
800	1,500	0.15	0.05	71.1	11.3	9.8	4.4	7.6	5.9
800	2,000	0.15	0.05	53.3	15.0	10.5	4.9	9.1	6.8
400	1,000	0.2	0.05	40.0	10.0	10.2	4.3	9.1	6.2
400	1,500	0.2	0.05	26.7	15.0	9.9	4.2	8.2	5.2
400	2,000	0.2	0.05	20.0	20.0	9.4	4.2	8.2	5.4
600	1,000	0.2	0.05	60.0	10.0	11.3	4.8	6.7	4.7
600	1,500	0.2	0.05	40.0	15.0	13.4	6.5	7.8	6.0
600	2,000	0.2	0.05	30.0	20.0	13.5	6.6	7.3	5.5
800	1,000	0.2	0.05	80.0	10.0	11.3	6.0	7.0	5.2
800	1,500	0.2	0.05	53.3	15.0	12.8	7.4	8.2	6.1
800	2,000	0.2	0.05	40.0	20.0	13.4	8.0	8.7	6.6
600	2,500	0.15	0.05	32.0	18.8	15.2	4.8	14.7	5.2
600	3,000	0.15	0.05	26.7	22.5	15.9	5.5	12.3	4.9
700	2,500	0.15	0.05	37.3	18.8	18.2	5.8	12.1	5.2

Laser Power [W]	Scan Speed [mm/s]	Hatch-Distance [mm]	Layer thickness [mm]	VED [J/mm³]	Build Rate [mm³/s]	Amount of Spatters for Gaussian Beam Profile		Amount of Spatters for Donut Beam Profile	
						AVG	STD	AVG	STD
700	3,000	0.15	0.05	31.1	22.5	16.3	5.3	12.1	4.5
800	2,500	0.15	0.05	42.7	18.8	19.4	7.0	11.2	4.2
800	3,000	0.15	0.05	35.6	22.5	17.3	5.4	13.1	5.4
900	2,500	0.15	0.05	48.0	18.8	17.8	6.5	11.5	7.1
900	3,000	0.15	0.05	40.0	22.5	19.4	7.9	11.6	5.0
600	2,500	0.2	0.05	24.0	25.0	16.3	5.0	12.9	5.6
600	3,000	0.2	0.05	20.0	30.0	17.5	7.2	10.9	5.3
700	2,500	0.2	0.05	28.0	25.0	19.9	6.6	10.5	5.0
700	3,000	0.2	0.05	23.3	30.0	20.9	10.1	10.2	4.6
800	2,500	0.2	0.05	32.0	25.0	24.5	13.3	12.1	6.6
800	3,000	0.2	0.05	26.7	30.0	24.1	10.8	15.1	6.6
900	2,500	0.2	0.05	36.0	25.0	21.3	9.6	12.9	5.7
900	3,000	0.2	0.05	30.0	30.0	20.6	10.8	12.3	5.5
600	2,500	0.25	0.05	19.2	31.3	10.6	5.2	11.6	5.4
600	3,000	0.25	0.05	16.0	37.5	11.9	5.7	9.6	4.4
700	2,500	0.25	0.05	22.4	31.3	14.4	7.5	9.9	4.7
700	3,000	0.25	0.05	18.7	37.5	14.9	7.6	10.1	4.3
800	2,500	0.25	0.05	25.6	31.3	16.1	8.3	13.3	6.3
800	3,000	0.25	0.05	21.3	37.5	16.5	9.4	13.0	6.5
900	2,500	0.25	0.05	28.8	31.3	19.2	9.6	12.9	6.5
900	3,000	0.25	0.05	24.0	37.5	19.9	10.3	10.4	5.0
			Mean Value			**15.2**	**6.7**	**9.8**	**5.3**
			Minimum Value			**9.4**	**4.2**	**5.4**	**3.5**
			Maximum Value			**24.5**	**13.3**	**15.1**	**7.1**

A.5 Single Track experiments

Table A-3 and Table A-4 provide an overview of the measured and simulated width and depth of the single tracks. The results are used for the validation of the numerical model (see Chapter 3.4) and to investigate the effect of the beam profile on the formation of single tracks (see Chapter 5.2).

Table A-3: Results of single-track experiments and simulation for Gaussian beam profile.

Laser Power [W]	Scan Speed [mm/s]	LED [J/mm]	Experimental Results: Width of Single Tracks [µm]		Depth of Single Tracks [µm]		Width / Depth	Numerical Results: Width of Single Tracks [µm]	Depth of Single Tracks [µm]	Width / Depth
			AVG	STD	AVG	STD	AVG			
400	1,000	0.40	253	19	68	240	3.6	240	84	2.9
400	1,500	0.27	187	15	44	215	5.2	215	61	3.5
400	2,000	0.20	151	21	38	185	4.8	185	43	4.3
400	2,500	0.16	147	9	31	165	6.4	165	38	4.3
500	1,000	0.50	274	10	89	258	3.1	258	116	2.2
500	1,500	0.33	215	11	57	241	3.8	241	85	2.8
500	2,000	0.25	195	13	43	215	4.6	215	60	3.6
500	2,500	0.20	197	14	41	193	4.8	193	47	4.1
600	1,000	0.60	325	26	108	293	3.0	293	160	1.8
600	1,500	0.40	242	20	81	268	3.0	268	103	2.6
600	2,000	0.30	214	15	57	245	3.8	245	82	3.0
600	2,500	0.24	202	21	50	215	4.0	215	59	3.6
700	1,000	0.70	342	55	167	318	2.1	318	190	1.7
700	1,500	0.47	314	42	105	301	3.0	301	129	2.3
700	2,000	0.35	235	16	67	275	3.5	275	93	3.0
700	2,500	0.28	235	18	62	245	3.8	245	70	3.5

Table A-4: Results of single-track experiments and simulation for donut beam profile.

Laser Power [W]	Scan Speed [mm/s]	LED [J/mm]	Experimental Results					Numerical Results		
			Width of Single Tracks [µm]		Depth of Single Tracks [µm]		Width / Depth	Width of Single Tracks [µm]	Depth of Single Tracks [µm]	Width / Depth
			AVG	STD	AVG	STD	AVG			
400	1,000	0.40	240	28	54	12	4.3	245	70	3.5
400	1,500	0.27	216	10	57	10	4.8	214	53	4.0
400	2,000	0.20	169	11	44	10	4.6	187	40	4.7
400	2,500	0.16	164	14	32	8	5.3	162	25	6.5
500	1,000	0.50	261	22	74	10	3.5	272	87	3.1
500	1,500	0.33	234	6	57	7	4.1	245	70	3.5
500	2,000	0.25	193	27	44	8	4.4	216	53	4.1
500	2,500	0.20	160	9	32	2	5.1	194	40	4.9
600	1,000	0.60	276	24	83	12	3.3	303	118	2.6
600	1,500	0.40	244	10	63	6	3.9	275	85	3.2
600	2,000	0.30	210	40	55	15	3.8	245	61	4.0
600	2,500	0.24	210	9	53	12	4.0	221	58	3.8
700	1,000	0.70	332	39	120	15	2.8	327	158	2.1
700	1,500	0.47	283	26	82	3	3.4	305	103	3.0
700	2,000	0.35	239	20	64	14	3.7	276	74	3.7
700	2,500	0.28	210	28	53	5	3.9	245	56	4.4

A.6 Wilcoxon-Signed-Rank-Test

The Wilcoxon-Signed-Rank-Test is a non-parametric statistical test that is used to determine whether or not two independent samples differ from each other and to calculate the confidence level of the underlying hypothesis [199, 217]. The test is done for the verification of the hypothesis that the beam profile has a statistically relevant effect on the melt pool width and depth of the single track. A detailed description of the test can be found in the literature [199, 217, 220].

The test is done for LEDs < 0.4 J/mm and LEDs ≥ 0.4 J/mm independently as there is a clear trend-shift above and below this value, especially for the melt pool depth (see Figure 5-19 on 81). Table A-5 provides an overview of the Wilcoxon-test for LEDs < 0.4 J/mm. The critical value k_W that is necessary for the analysis can be looked up in corresponding tables and is defined as $k_W = 8$ for a sample size of $n = 10$ and $k_W = 0$ for a sample size of $n = 6$ for a confidence level of 95% [199]. As the minimum value of the sum ranks is "SUM-" = 0 for the melt pool depth at LEDs ≥ 0.4 J/mm, the assumption that there is a statistically relevant effect of the beam profile on the melt pool depth is confirmed for LEDs ≥ 0.4 J/mm. The effect of the beam profile on the melt pool width on the other hand is not statistically relevant based on a confidence level of 95%.

Table A-5: Data of Wilcoxon-Test for LEDs < 0.4 J/mm

			Width of Single Tracks [µm]					Depth of Single Tracks [µm]				
Laser Power [W]	**Scan Speed [mm/s]**	**LED [J/mm]**	**Gaussian Profile**	**Donut Profile**	**Sign**	**Value**	**Rank**	**Gaussian**	**Donut**	**Sign**	**Value**	**Rank**
400	1,500	0.27	187	216	-1	29.1	-13	44	57	-1	12.5	-10
400	2,000	0.20	151	169	-1	18.1	-10	38	44	-1	6.2	-7
400	2,500	0.16	147	164	-1	16.7	-9	31	32	-1	1.0	-2
500	1,500	0.33	215	234	-1	18.6	-11	57	57	1	0.5	1
500	2,000	0.25	195	193	1	2.4	1	43	44	-1	1.4	-3
500	2,500	0.20	197	160	1	37.0	15	41	32	1	9.3	9
600	2,000	0.30	214	210	1	3.8	3	57	55	1	1.6	4
600	2,500	0.24	202	210	-1	7.8	-5	50	53	-1	3.0	-5.5
700	2,000	0.35	235	239	-1	4.8	-4	67	64	1	3.0	5.5
700	2,500	0.28	235	210	1	25.0	12	62	53	1	8.3	8
					SUM Rank +		**31**			**SUM Rank +**		**27.5**
					SUM Rank -		**52**			**SUM Rank -**		**27.5**
						n	**10**				**n**	**10**

Table A-6: Data of Wilcoxon-Test for LEDs ≥ 0.4 J/mm

Laser Power [W]	Scan Speed [mm/s]	LED [J/mm]	Width of Single Tracks [µm]					Depth of Single Tracks [µm]				
			Gaussian Profile	Donut Profile	Sign	Value	Rank	Gaussian	Donut	Sign	Value	Rank
400	1,000	0.40	253	240	1	13.8	8	68	54	1	14.4	11
500	1,000	0.50	274	261	1	13.8	7	89	74	1	15.6	12
600	1,000	0.60	325	276	1	49.5	16	108	83	1	25.7	15
600	1,500	0.40	242	244	-1	2.8	-2	81	63	1	18.2	13
700	1,000	0.70	342	332	1	10.4	6	167	120	1	46.6	16
700	1,500	0.47	314	283	1	31.4	14	105	82	1	22.2	14
						SUM +	**51**				**SUM +**	**81**
						SUM -	**2**				**SUM -**	**0**
						n	**6**				**n**	**6**

A.7 Micro-sections of AlSi10Mg density cubes

The following tables give an overview of representative micro-sections of the density cubes manufactured in AlSi10Mg.

Table A-7: Comparison of micro-sections for different scanning speeds and laser powers for a hatch distance of 100 µm for AlSi10Mg (G = Gaussian beam profile; D = donut beam profile).

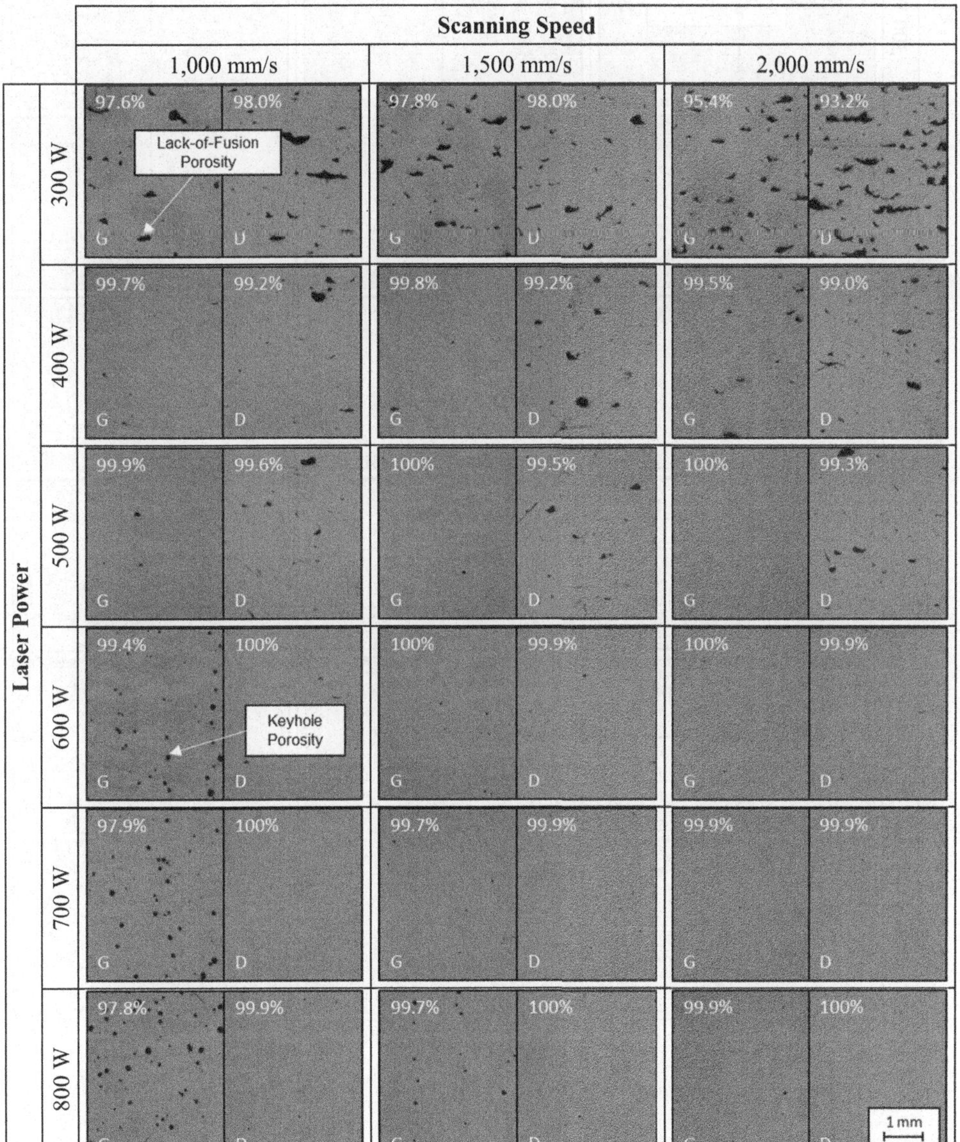

Table A-8: Comparison of micro-sections for different scanning speeds and laser powers for a hatch distance of 150 µm for AlSi10Mg (G = Gaussian beam profile; D = donut beam profile).

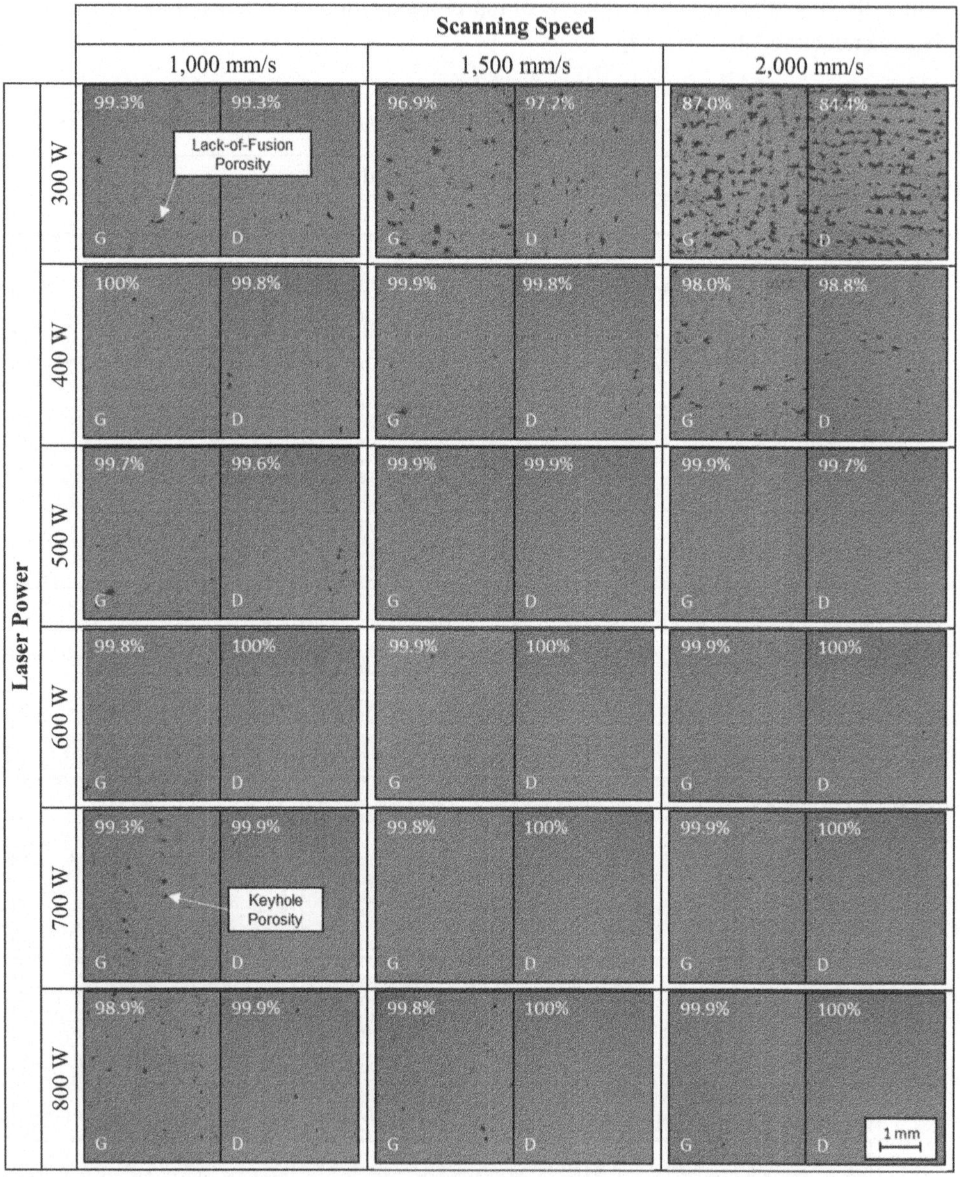

Table A-9: Comparison of micro-sections for different scanning speeds and laser powers for a hatch distance of 150 μm for AlSi10Mg (G = Gaussian beam profile; D = donut beam profile).

		Scanning Speed	
		2,500 mm/s	3,000 mm/s
Laser Power	600 W	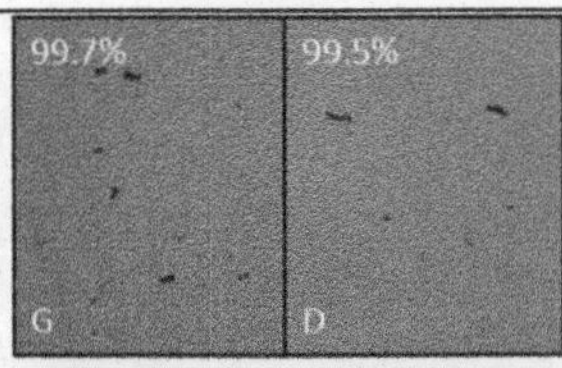	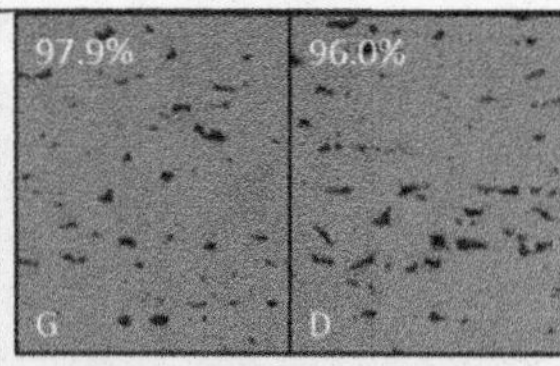
	700 W	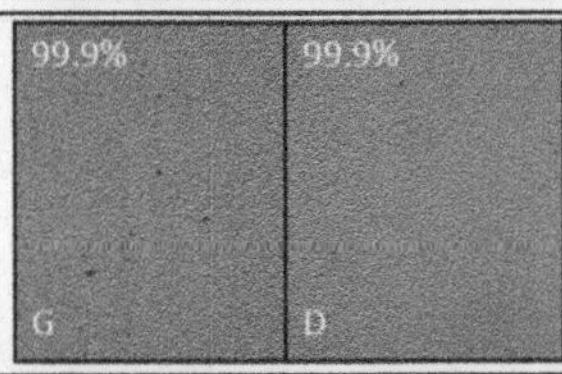	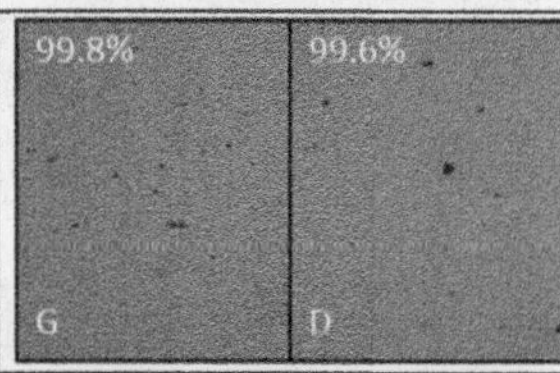
	800 W	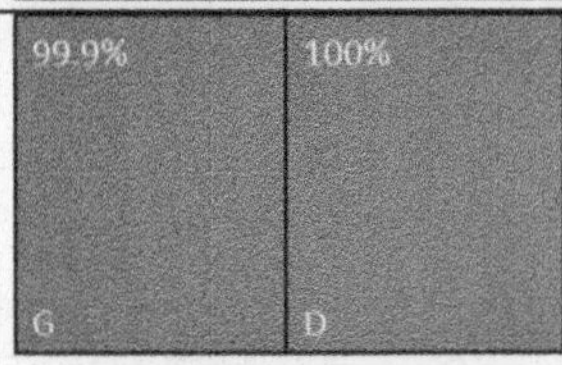	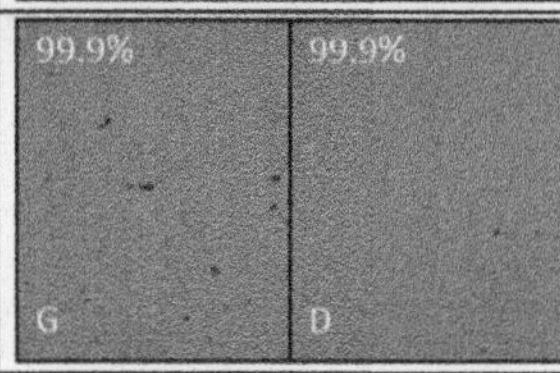
	900 W	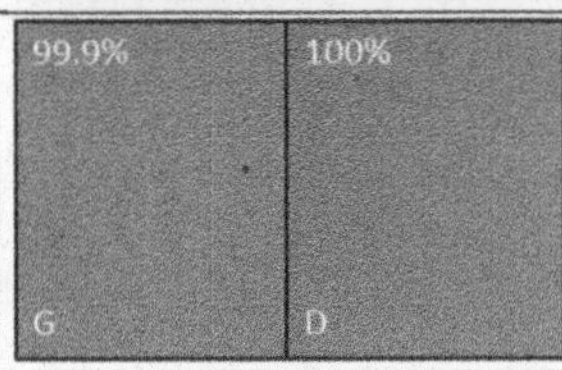	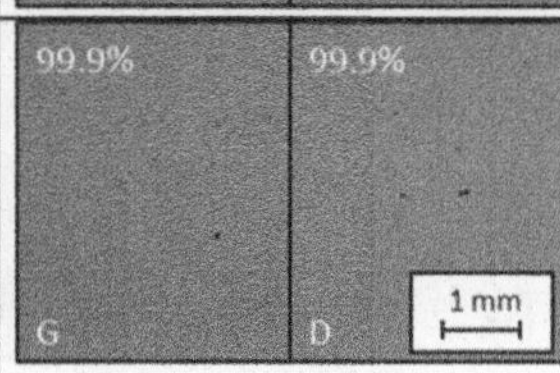

Table A-10: Comparison of micro-sections for different scanning speeds and laser powers for a hatch distance of 200 µm for AlSi10Mg (G = Gaussian beam profile; D = donut beam profile).

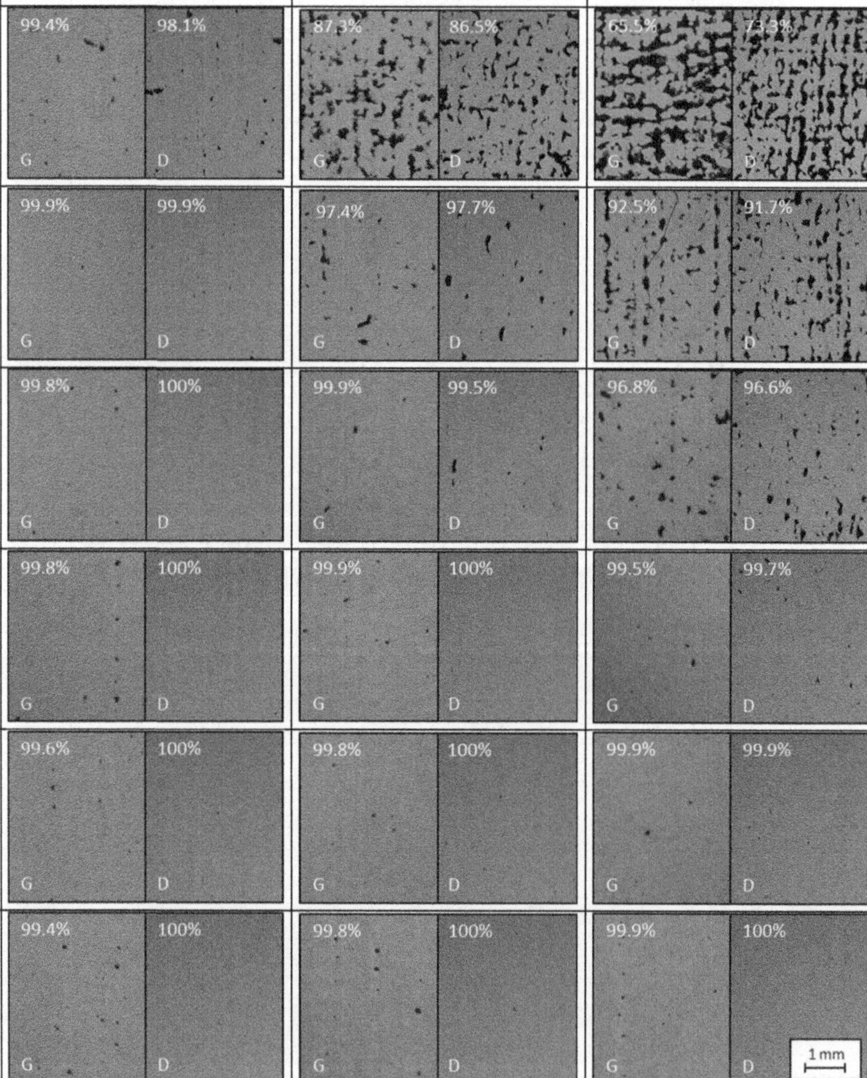

Laser Power	Scanning Speed 1,000 mm/s		1,500 mm/s		2,000 mm/s	
	G	D	G	D	G	D
300 W	99.4%	98.1%	87.3%	86.5%	65.5%	73.3%
400 W	99.9%	99.9%	97.4%	97.7%	92.5%	91.7%
500 W	99.8%	100%	99.9%	99.5%	96.8%	96.6%
600 W	99.8%	100%	99.9%	100%	99.5%	99.7%
700 W	99.6%	100%	99.8%	100%	99.9%	99.9%
800 W	99.4%	100%	99.8%	100%	99.9%	100%

Table A-11: Comparison of micro-sections for different scanning speeds and laser powers for a hatch distance of 200 µm for AlSi10Mg (G = Gaussian beam profile; D = donut beam profile).

		Scanning Speed	
		2,500 mm/s	3,000 mm/s
Laser Power	600 W	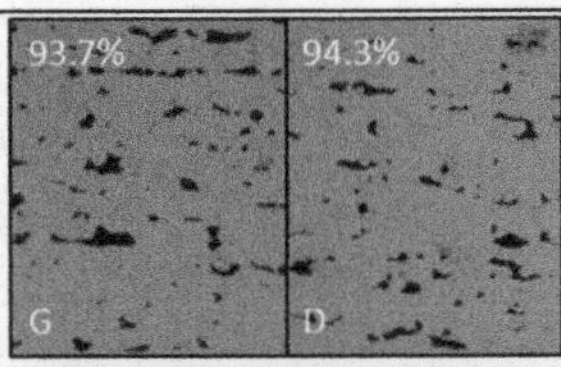	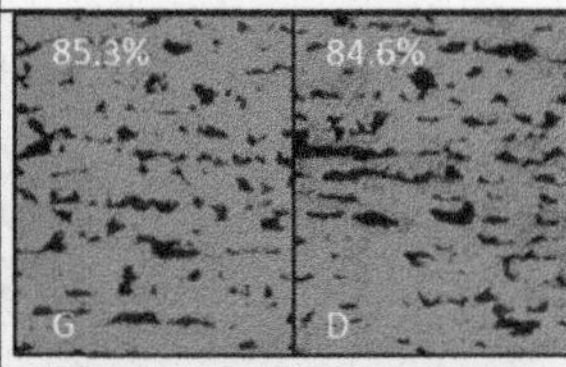
	700 W	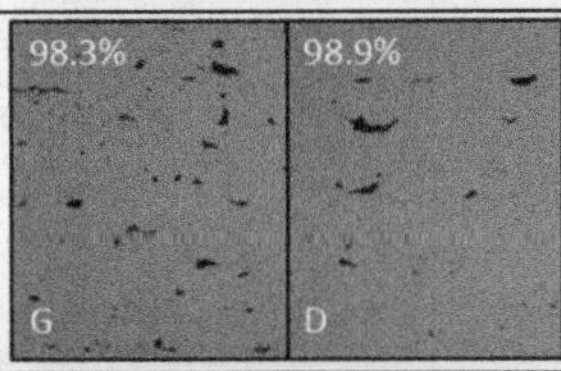	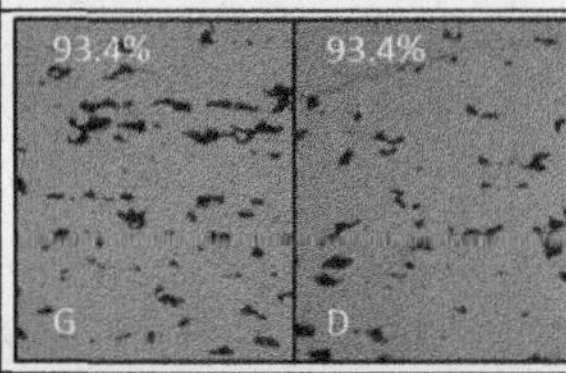
	800 W	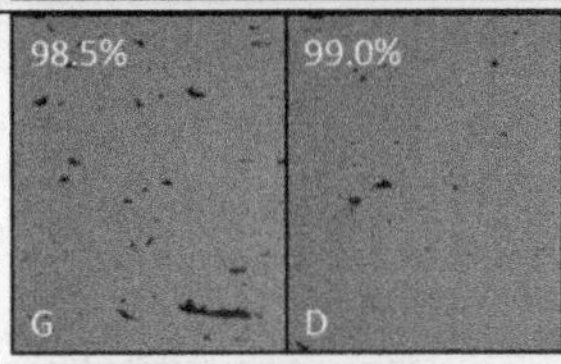	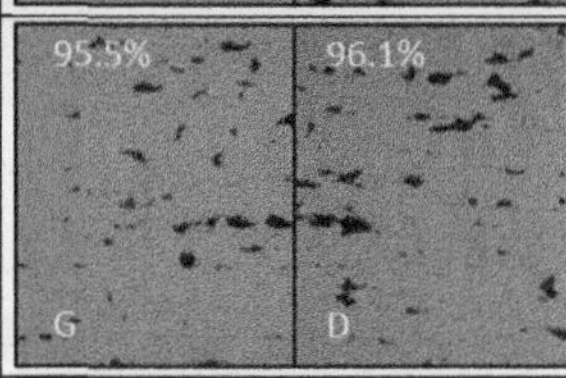
	900 W	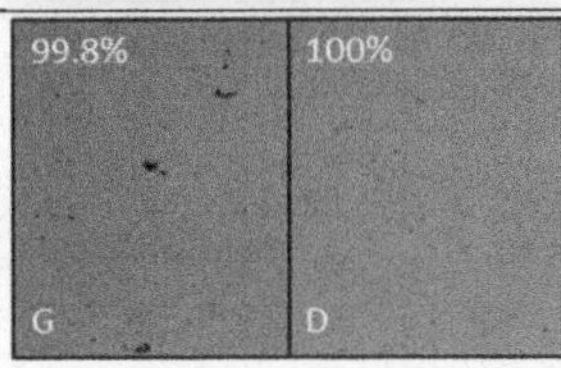	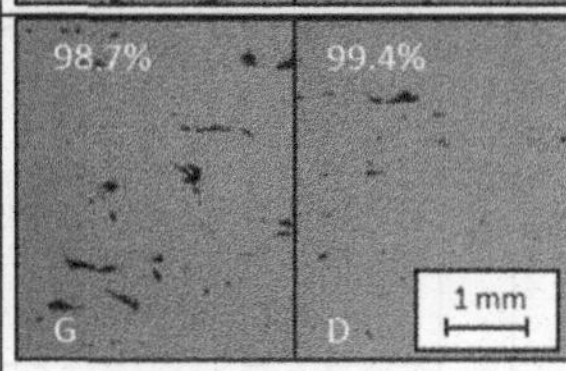

Table A-12: Comparison of micro-sections for different scanning speeds and laser powers for a hatch distance of 250 µm for AlSi10Mg (G = Gaussian beam profile; D = donut beam profile).

		Scanning Speed	
		2,500 mm/s	3,000 mm/s
Laser Power	600 W	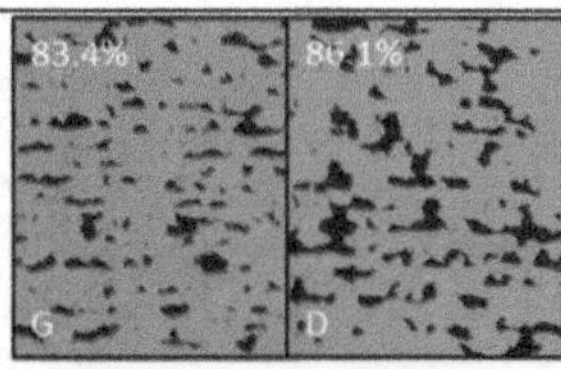	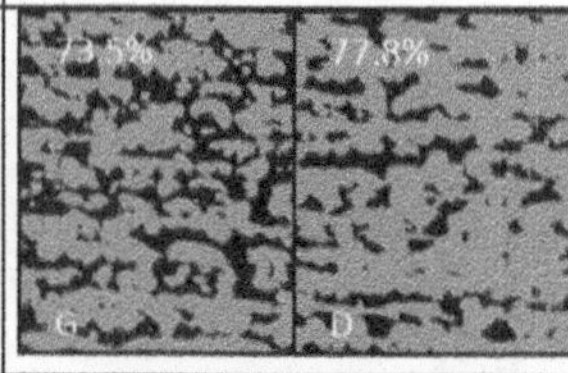
	700 W	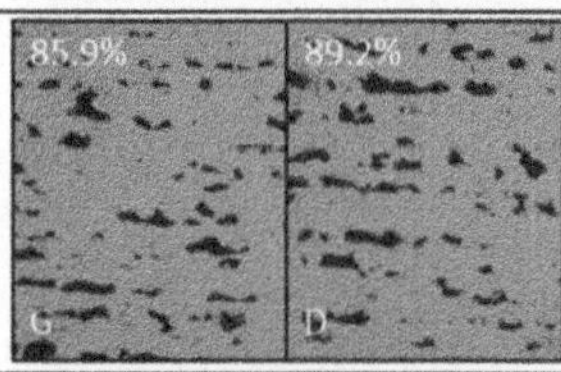	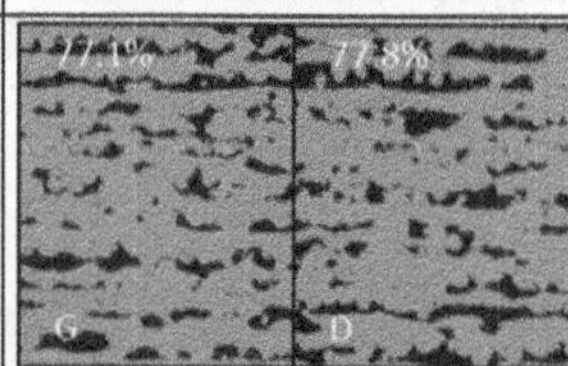
	800 W	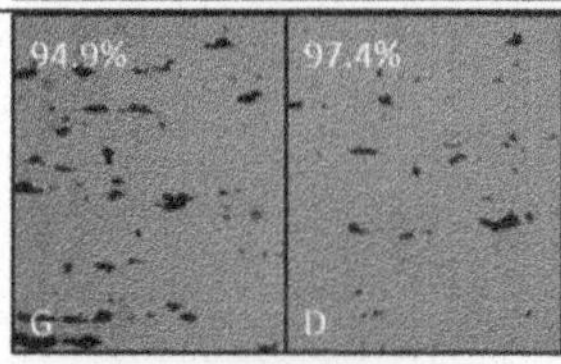	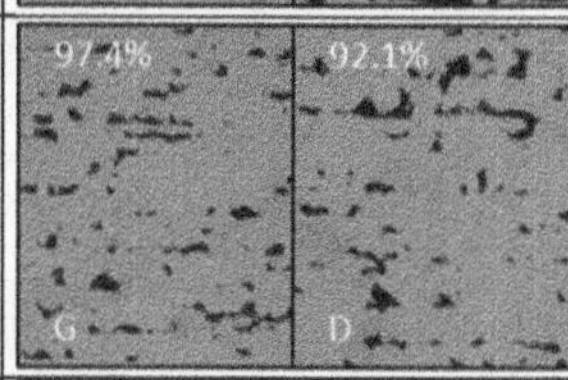
	900 W	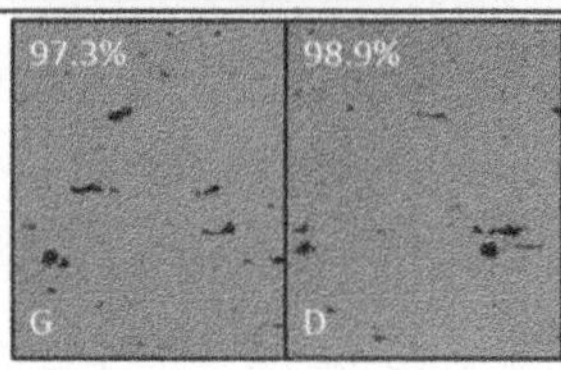	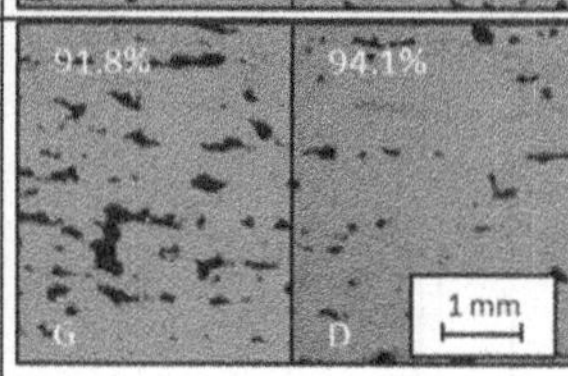

A.8 Micro-sections of AlSi16Sc density cubes

The following tables provide an overview of representative micro-sections of the density cubes manufactured in AlSi16Sc.

Table A-13: Comparison of micro-sections for different scanning speeds and laser powers for a hatch distance of 100 µm for AlSi16Sc (G = Gaussian beam profile; D = donut beam profile).

Laser Power		Scanning Speed 1,000 mm/s G	1,000 mm/s D	1,500 mm/s G	1,500 mm/s D	2,000 mm/s G	2,000 mm/s D
Laser Power	300 W	98.4%	99.5%	98.5%	99.7%	95.8%	99.2%
	400 W	99.9%	99.9%	99.9%	99.9%	99.7%	99.9%
	500 W	100%	99.9%	99.9%	100%	100%	100%
	600 W	97,5%	100%	99.9%	100%	99.9%	99.9%
	700 W	95.1%	99.8%	98.0%	100%	99.6%	100%
	800 W	94.4%	99.3%	96.6%	99.9%	98.8%	99.9%

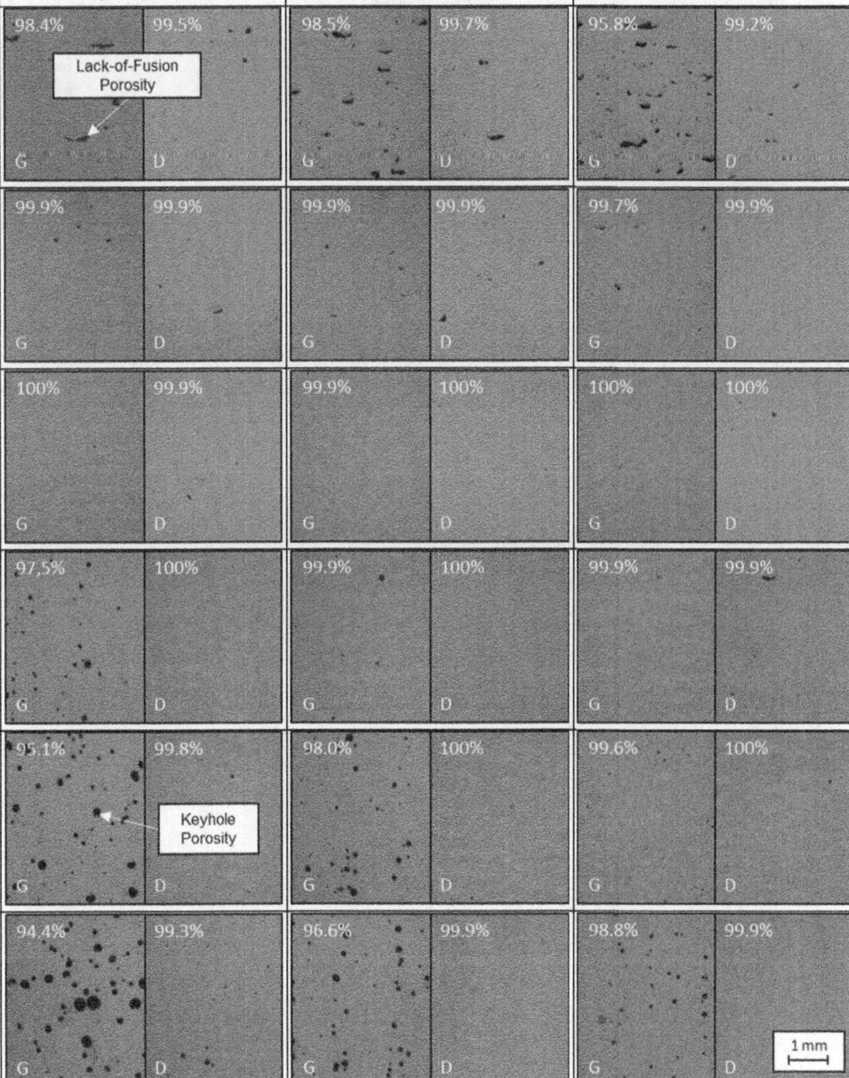

Table A-14: Comparison of micro-sections for different scanning speeds and laser powers for a hatch distance of 150 µm for AlSi16Sc (G = Gaussian beam profile; D = donut beam profile).

Laser Power	Scanning Speed 1,000 mm/s G	1,000 mm/s D	1,500 mm/s G	1,500 mm/s D	2,000 mm/s G	2,000 mm/s D
300 W	95.8%	98.6%	96.5%	99.1%	86.4%	93.6%
400 W	99.9%	99.7%	99.9%	99.7%	99.1%	98.4%
500 W	99.6%	99.8%	99.9%	99.9%	99.9%	99.9%
600 W	97.7%	99.9%	99.8%	99.9%	99.9%	100%
700 W	95.8%	99.9%	99.4%	100%	99.6%	100%
800 W	94.9%	99.9%	98.4%	99.9%	99.2%	99.9%

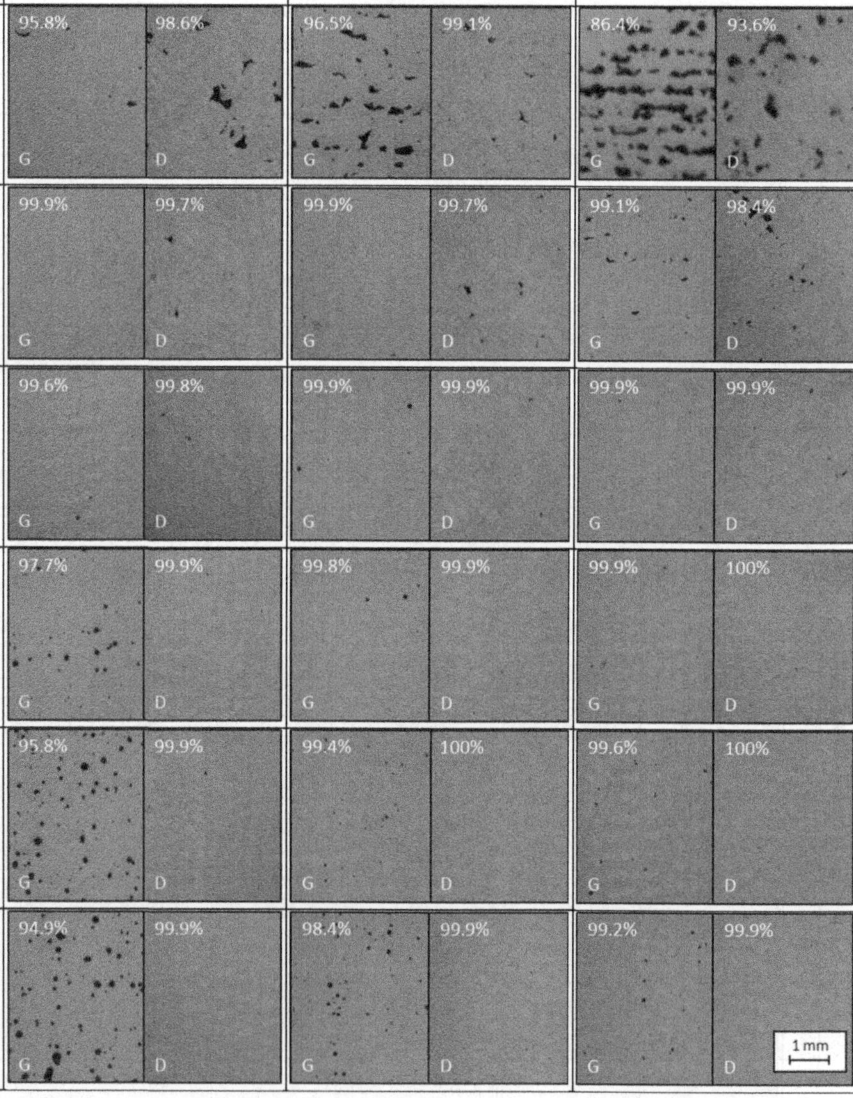

Table A-15: Comparison of micro-sections for different scanning speeds and laser powers for a hatch distance of 200 µm for AlSi16Sc (G = Gaussian beam profile; D = donut beam profile).

		Scanning Speed					
		1,000 mm/s		1,500 mm/s		2,000 mm/s	
Laser Power	300 W	99.0% G	99.4% D	88.2% G	88.7% D	70.9% G	82.9% D
	400 W	99.9% G	99.7% D	99.4% G	99.9% D	91.4% G	94.9% D
	500 W	99.8% G	99.7% D	99.9% G	99.9% D	97.9% G	97.5% D
	600 W	99.4% G	99.9% D	99.7% G	99.9% D	99.8% G	99.9% D
	700 W	96.6% G	99.8% D	99.4% G	99.8% D	99.5% G	99.8% D
	800 W	96.5% G	99.7% D	99.2% G	99.8% D	99.6% G	99.9% D

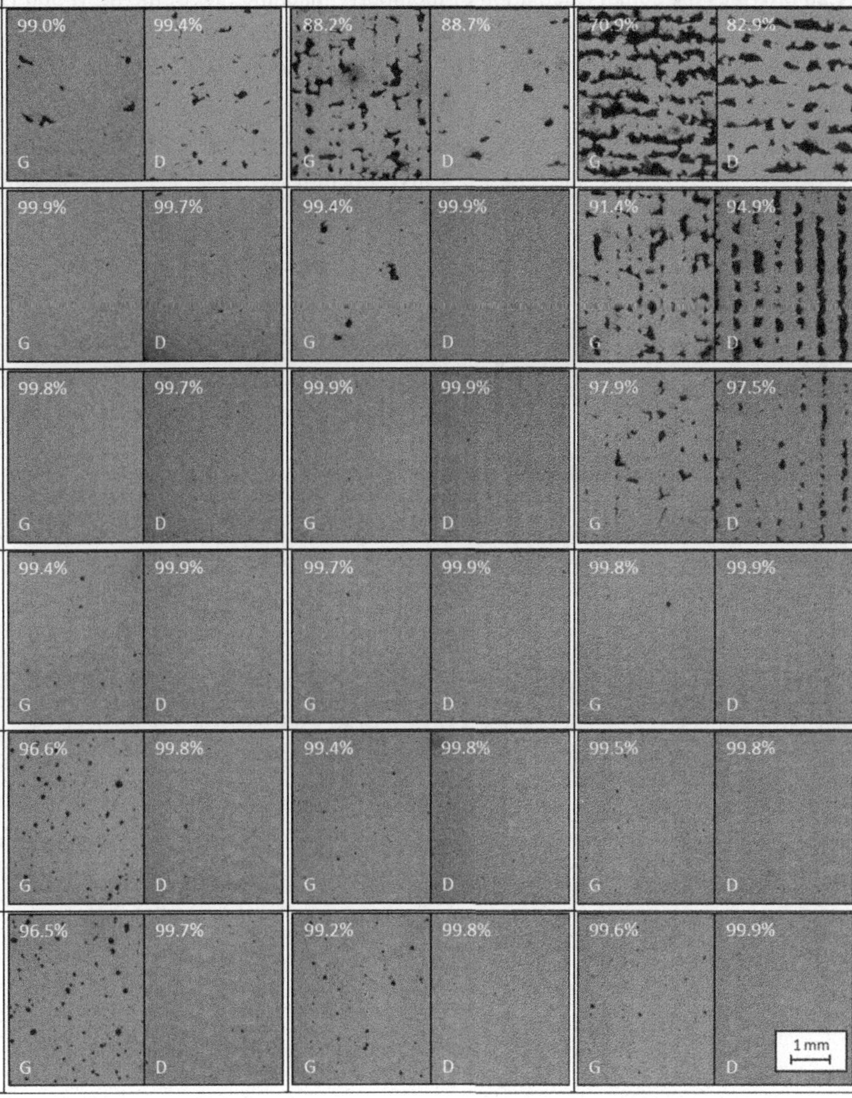

A.9 Hourly Machine Rate of SLM 500 HL Machine

In the following, the calculation of the hourly machine rate of the SLM 500 HL is laid out. Figure A-9 gives an overview of the key data that are used for the calculation and Equation A.28 provides the actual calculation of the hourly machine rate of the SLM 500 HL. The calculation only considers the key cost drivers and neglects parameters with a minor effect such as electricity or gas consumption [202]. With the provided data, an hourly rate of $C_h = 71$ €/h is the calculated result.

- Investment C_{inv}: 1,200,000 €
- Depreciation time t_{Dep}: 5 years
- Occupancy rate a_{Occ}: 65%
- Maintenance costs per year C_m: 30,000 €/a
- Overhead costs C_{oh}: 50%

Figure A-9: Key data for calculation of the hourly machine rate of SLM 500 HL.

$$C_h = \left(\frac{C_{inv}}{t_{dep} \cdot a_{occ}} + \frac{C_m}{o_{occ}}\right) C_{oh} = \left(\frac{1{,}200{,}000\,€}{5 \cdot 365 \cdot 24h \cdot 0.65} + \frac{30{,}000\,€}{365 \cdot 24h \cdot 0.65}\right) 1.5 = 71\frac{€}{h} \quad \text{(A.28)}$$

C_h:	Hourly machine rate of SLM 500 HL [€/h]	C_{inv}:	Investment costs for SLM 500 [€]
C_m:	Maintenance costs per year [€/year]	t_{dep}:	Depreciation time of machine [year]
a_{occ}:	Occupancy rate of the machine [%]	C_{oh}:	Overhead costs [%]

The manufacturer's authorised representative in the EU is Springer Nature Customer Service Centre GmbH, Europaplatz 3, 69115 Heidelberg, Germany. If you have any concerns regarding our products, please contact ProductSafety@springernature.com

Printed and bound by CPI Group (UK) Ltd, Croydon, CR0 4YY
30/09/2026
02225235-0013